T0271977

An Introduction to
Radiation Protection in Medicine

Series in Medical Physics and Biomedical Engineering

Series Editors: John G Webster, E Russell Ritenour, Slavik Tabakov, and Kwan-Hoong Ng

Series in Medical Physics and Biomedical Engineering

An Introduction to Radiation Protection in Medicine

Edited by

Jamie V. Trapp
Queensland University of Technology
Brisbane, Australia

Tomas Kron
Peter MacCallum Cancer Centre
Melbourne, Australia

Taylor & Francis
Taylor & Francis Group
New York   London

Taylor & Francis is an imprint of the
Taylor & Francis Group, an **informa** business

CRC Press
Taylor & Francis Group
6000 Broken Sound Parkway NW, Suite 300
Boca Raton, FL 33487-2742

International Standard Book Number-13: 978-1-58488-964-9 (Hardcover)

Library of Congress Cataloging-in-Publication Data

An introduction to radiation protection in medicine / editors, Jamie V. Trapp and Tomas Kron.
    p. ; cm. -- (Series in medical physics and biomedical engineering)
    Includes bibliographical references and index.
    ISBN 978-1-58488-964-9 (hardcover : alk. paper) 1. Radiology, Medical--Safety measures. 2. Radiation--Safety measures. I. Trapp, Jamie V. II. Kron, Tomas. III. Series.
    [DNLM: 1. Radiation Protection--methods. 2. Radiation Injuries--prevention & control. 3. Radiography--adverse effects. 4. Radiology. 5. Radiotherapy--adverse effects. WN 650 I6955 2008]

R895.I72 2008
363.17'90289--dc22
                                                                     2007041279

Visit the Taylor & Francis Web site at
http://www.taylorandfrancis.com

and the CRC Press Web site at
http://www.crcpress.com

Contents

About the Series

The Series in Medical Physics and Biomedical Engineering describes the applications of physical sciences, engineering, and mathematics in medicine and clinical research.

The series seeks (but is not restricted to) publications in the following topics:

- Artificial organs
- Assistive technology
- Bioinformatics
- Bioinstrumentation
- Biomaterials
- Biomechanics
- Biomedical engineering
- Clinical engineering
- Imaging
- Implants
- Medical computing and mathematics
- Medical/surgical devices
- Patient monitoring
- Physiological measurement
- Prosthetics
- Radiation protection, health physics, and dosimetry
- Regulatory issues
- Rehabilitation engineering
- Sports medicine
- Systems physiology
- Telemedicine
- Tissue engineering
- Treatment

The Series in Medical Physics and Biomedical Engineering is an international series that meets the need for up-to-date texts in this rapidly developing field. Books in the series range in level from introductory graduate textbooks and practical handbooks to more advanced expositions of current research.

The Series in Medical Physics and Biomedical Engineering is the official book series of the International Organization for Medical Physics.

The International Organization for Medical Physics

The International Organization for Medical Physics (IOMP), founded in 1963, is a scientific, educational, and professional organization of 76 national adhering organizations, more than 16,500 individual members, several corporate members, and four international regional organizations.

IOMP is administered by a council, which includes delegates from each of the adhering national organizations. Regular meetings of the council are held electronically as well as every three years at the World Congress on Medical Physics and Biomedical Engineering. The president and other officers form the executive committee, and there are also committees covering the main areas of activity, including education and training, scientific, professional relations, and publications.

Objectives

- To contribute to the advancement of medical physics in all its aspects
- To organize international cooperation in medical physics, especially in developing countries
- To encourage and advise on the formation of national organizations of medical physics in those countries which lack such organizations

Activities

Official journals of the IOMP are *Physics in Medicine and Biology* and *Medical Physics and Physiological Measurement*. The IOMP publishes a bulletin *Medical Physics World* twice a year, which is distributed to all members.

A World Congress on Medical Physics and Biomedical Engineering is held every three years in cooperation with IFMBE through the International Union for Physics and Engineering Sciences in Medicine (IUPESM). A regionally based international conference on medical physics is held between world congresses. IOMP also sponsors international conferences, workshops, and courses. IOMP representatives contribute to various international committees and working groups.

The IOMP has several programs to assist medical physicists in developing countries. The joint IOMP Library Programme supports 69 active libraries in 42 developing countries, and the Used Equipment Programme coordinates equipment donations. The Travel Assistance Programme provides a limited number of grants to enable physicists to attend the world congresses. The IOMP Web site is being developed to include a scientific database of international standards in medical physics and a virtual education and resource center.

Information on the activities of the IOMP can be found on its Web site at www.iomp.org.

Preface

This book is the result of informal meetings of a group of medical physicists in Melbourne, Australia. We all share an interest in teaching and radiation protection, but have very different roles in various hospitals and universities. Although this book is written for all health professionals who work with radiation, it originally arose from discussions about the need to equip medical physics trainees with knowledge about radiation safety in a fast-changing and diverse environment such as medicine.

From the very start it was meant to be a practical book. Half of it is concerned with basic issues related to radiation protection, while the other half addresses situations commonly encountered in the three major areas in medicine where radiation is the tool to diagnose or treat human disease: radiology, nuclear medicine, and radiation therapy. It is interesting to note that these fields are becoming more entangled—from positron emission tomography–computed tomography (PET-CT) scanners to image-guided radiotherapy, radiation protection professionals are concerned with a multitude of different issues. The present book aims to equip people with the background to function in such a versatile and fast-developing field.

While the book should be seen as a whole, we did not try to make all chapters sound the same. By maintaining the individual style of each of the authors we are hoping to make accessible their vast personal experience.

On behalf of all the authors, thanks and gratitude are extended to the various spouses, family members, and friends, who bore with this group of "grumpy old men" during the process of preparing the book.

We hope you enjoy the reading and find it interesting and relevant. Please provide us with feedback and suggestions about the book—after all, learning never ends.

Tomas Kron and Jamie Trapp

July 2007

Editors/Contributors

Jamie Trapp has several years of research, clinical, and teaching experience in medical and radiation physics. His current post is lecturer in radiation physics at Queensland University of Technology, and he was previously a senior lecturer in medical physics at Royal Melbourne Institute of Technology, a research scientist at the Institute of Cancer Research in London, and a medical physicist at the Princess Alexandra Hospital in Brisbane, Australia. Before starting a career in physics he held several positions, including spending 6 years enlisted as a clerk in the Royal Australian Air Force. He is a member of the Australasian Radiation Protection Society and the Institute of Physics. His research interests include novel detector development and radiation dosimetry.

Tomas Kron is a radiotherapy physicist with more than 20 years of experience in medical physics. He holds accreditations in radiotherapy physics from the Canadian College of Physicists in Medicine (CCPM) and the Australasian College of Physical Scientists and Engineers in Medicine (ACPSEM) and is certified for radiation protection by the Australasian Radiation Protection Accreditation Board (ARPAB). Tomas has an interest in image-guided radiation therapy, clinical dosimetry, and the education of medical physicists.

Raymond Budd is a senior lecturer (sessional) in undergraduate and postgraduate degree courses in medical radiation sciences at Monash University. He is also a radiation protection physicist with more than 30 years of experience in medical physics. He holds accreditation in radiological physics from the Australasian College of Physical Scientists and Engineers in Medicine (ACPSEM) and is certified for radiation protection by the Australasian Radiation Protection Accreditation Board (ARPAB). He has an interest in all aspects of radiation safety and patient dosimetry in diagnostic applications of radiation.

Jim Cramb has over 35 years of experience in radiotherapy physics and radiation safety and has been director of the Physical Sciences Department at Peter MacCallum Cancer Institute for 7 years. He is accredited in radiotherapy physics by the Australasian College of Physical Scientists and Engineers in Medicine (ACPSEM). He has particular interests in radiation dosimetry and radiotherapy treatment planning, including intensity-modulated radiotherapy (IMRT).

Ram Das is a radiotherapy physicist with more than 50 years of experience in radiation safety and medical physics. He is accredited as a radiotherapy physicist by the Australasian College of Physicists and Engineers in

Medicine. Formerly he was deputy director of the Atomic Energy Regulatory Board in India. He was also a member of the IAEA/WHO International Working Party on radium substitutes and afterloading techniques for cancer of the uterine cervix. Ram is interested in image-guided brachytherapy for prostate and gynecological cancers and in intraoperative brachytherapy.

John Heggie is a radiological physicist with more than 30 years of experience in medical physics and radiation safety. He holds accreditation in radiological physics from the Australasian College of Physical Scientists and Engineers in Medicine (ACPSEM). His interests are in the technology of medical imaging with particular emphasis on optimization in computed tomography (CT) and mammography.

Peter Johnston has been head of physics at Royal Melbourne Institute of Technology since 2003. He is a councillor of the Association of the Asia Pacific Physical Societies and of the Australian Institute of Nuclear Science and Engineering. He is a member of the Australian Governments Radiation Health and Safety Advisory Council and the Nuclear Safety Committee and a former member of the Radiation Health Committee. He is also an independent member of the Alligator Rivers Region Technical Committee. In 2006, he was a member of the Prime Ministers Uranium Mining and Nuclear Energy Review in Australia. Until February 2007, he was registrar and a member of the National Executive of the Australian Institute of Physics. Professor Johnston has had considerable experience in health and safety associated with environmental radioactivity, especially through his involvement with rehabilitation of the former British nuclear test site at Maralinga. His research interests include medical physics, health physics, environmental radioactivity, and ion beam physics.

1

Introduction

Tomas Kron

CONTENTS

1.1 Outline and Objective of the Book

The use of ionizing radiation has transformed medicine in the last 100 years. While excitement and curiosity dominated the first years, a realization about the risks of radiation for humans quickly ensued. One of the first reports of "traumatism from roentgen ray exposure" stems, from D. Walsh in 1897, just 2 years after the discovery of x-rays by C. W. Röntgen:[1] "So far from that view representing the exact state of affairs, it seems to the present writer that the method of roentgen ray diagnosis may exert a definitive harmful action upon some of the deeper tissues of the human body."[1]

The first formal approaches to limit exposure to ionizing radiation came into effect just before the First World War. Since then, there has always been an attempt to balance risks and benefit of the use of ionizing radiation in medicine, and it is probably fair to say that radiation safety has been a model for other fields where humans need to be protected from potentially harmful agents. The weighting of risks and benefits is not always an easy task, as different individuals tend to judge risk and benefit differently. What is an acceptable risk to one person (e.g., in exchange for a good diagnostic image) may be unacceptable to another person. The present book aims to contribute to this discussion by:

- Providing the philosophical and scientific background for radiation protection in medicine (Chapters 1 and 3).
- Discussing units and tools used in radiation protection (Chapters 2 and 4).
- Introducing concepts of a system of radiation protection in the workplace (Chapter 5).
- Giving detailed information on practical application of radiation protection in the three fields of medicine where radiation is most widely used: radiology (Chapter 6), nuclear medicine (Chapter 7), and radiotherapy (Chapters 8 and 9); the latter requires two chapters as the issues faced in external beam radiotherapy and brachytherapy are quite different.

After this overview, the present introductory chapter will introduce the framework of radiation protection introduced by the International Commission on Radiological Protection (ICRP). The 1990 recommendations published in ICRP Report 60[2] build the foundation of most radiation protection regulations in the world. Whilst new recommendations have been published as ICRP Report 103[12], the recommendations will take years to be promulgated through to national regulations; thus this discussion is based on ICRP Report 60. A preview of the new recommendations can be found also in two papers by R. Clarke, the long-time chairman of the ICRP.[3,4]

This is followed in Chapter 2 by a summary of the background of physical science behind radiation protection. Chapter 3 provides the radiobiological basis of radiation protection, which can help the reader to appreciate the need for radiation protection as well as the order of magnitude chosen for limit values and constraints. Readers with a radiological physics or radiobiology background may be able to skip Chapters 2 and 3, respectively.

Chapters 4 and 5 introduce tools for radiation protection; Chapter 4 discusses instrumentation while Chapter 5 focuses on operational and managerial tools for organizing radiation safety in a medical workplace.

The last four chapters on the practice of radiation protection in different medical disciplines provide the core of the material presented. They are informed by experience and as such written with a focus on practical issues facing persons concerned with radiation safety in a hospital. These application chapters are supported by background information in the earlier chapters and a glossary that explains terms used in the book.

This book should be of interest to all professionals working with radiation in medicine. It would be helpful to radiation protection professionals in medicine and to health physicists trying to familiarize themselves with the particular problems faced during the application of ionizing radiation in medicine. There is little assumed knowledge as Chapters 7 to 10 provide an introduction to the respective modalities used. As such, the book would also be relevant to students in all areas of science and medicine where ionizing radiation is used. This would also apply to junior doctors, radiographers, and nuclear medicine scientists.

In a teaching environment the material should be suitable to support a course at the senior undergraduate or early postgraduate level, with each chapter providing enough material for a 2-hour lecture plus associated laboratories and site visits.

The reader is also referred for updated information to the web pages of the International Commission on Radiological Protection (ICRP; http://www.icrp.org/index.asp) and the International Atomic Energy Agency (IAEA; http://www.iaea.org/); most of the technical documents are available as free downloads) including a special information page for patients on radiation in medicine (http://rpop.iaea.org/RPoP/RPoP/Content/index.htm). Also professional organizations such as the American Association of Physicists in Medicine (http://www.aapm.org/), the UK Institute of Physics and Engineering in Medicine (http://www.ipem.ac.uk/ipem%5Fpublic/), and the Health Physics Society (http://www.hps.org/) provide access to updated practical information for members and visitors to their sites.

1.2   Overview of the Use of Ionizing Radiation in Medicine

Many of the discoveries that advanced our understanding of human anatomy and function happened in the last few years of the nineteenth century. The discovery of x-rays by C. W. Roentgen in 1895 was a sensation—not the least because he included an x-ray image of the hand of his wife with the first reprints of his paper in the *Wuerzburger Medizinhistorischen Mitteilungen* sent to influential colleagues in the field.[5] The relevance for medicine was immediately realized and, very soon x-rays found applications on the battlefield in removing bullets and shrapnel from injured persons. Roentgen's discovery was followed by the discovery of radioactivity by H. Bequerel, who noted in 1896 that uranium was emitting radiation without the need of exposure to sunlight beforehand (as such it could not be fluorescence). In 1898, Marie Sklowdowska Curie announced the discovery of radium, which is much more radioactive than uranium.[6] The higher specific activity allowed the production of smaller sources that were useful in medicine for the treatment of a variety of lesions. Curie was instrumental in ensuring that her discovery found applications in helping others, for example, equipping mobile radiography units often referred to as petite Curies. Curie died in 1934 from aplastic anemia, a disease most likely due to her lifelong exposure to ionizing radiation.

Nowadays, ionizing radiation is used in medicine for two main purposes:

- Diagnostics
- Therapy

1.2.1   Diagnostics

The largest source of man-made radiation exposures to humans stems from diagnostic procedures. These can be broadly divided into radiology and nuclear medicine. This distinction is carried forward in the present book in Chapters 6 and 7.

In radiology, an external radiation source is used to generate photons (typically x-rays) of sufficient energy to penetrate human tissues. A detector system at the exit site of the beam determines the transmitted photons, which provide a projection image of all structures in the body. The contrast between adjacent tissues stems from their difference in electron density and atomic number (see Chapter 2). This information can be used as a single projection image (radiograph), for example, in chest x-rays. However, electronic detector systems also allow the acquisition of many consecutive images of the same anatomy. This fluoroscopic imaging allows the clinician to follow anatomical movement (e.g., breathing), surgical interventions (e.g., fluoroscopy-guided biopsy), or search for structures in the patient by moving the imaging apparatus. Finally, multiple projection images of the same anatomy can be acquired from different directions and a computer used to reconstruct three-dimensional information. This process, called computed tomography (CT), is usually done in several sections to reduce the influence of scatter on the reconstructed image.

In nuclear medicine, a radioactive isotope, most commonly 99m-technetium, is administered to the patient. The isotope is used to label a substance of interest that will follow a physiological pathway of interest. As such, nuclear medicine provides information about not just anatomical features in the patient but also metabolic activity and physiological pathways (often referred to as functional imaging). The emitted radiation can be detected from outside the patient using mostly gamma cameras, sophisticated arrangements of collimator and detector systems that allow determination of the location of the isotope. A method to improve spatial localization is the use of positron emitters, such as 18-fluorine. In this case, the positron will annihilate with an electron in the patient resulting in two 511 keV photons emitted approximately in opposing directions. Using coincidence measurements in a ring of detectors around the patient, it is possible to determine the exact trajectory on which the original positron interaction has taken place. Positron emission tomography (PET) has recently had a tremendous surge of interest as a very important oncological imaging modality. It is also a good example for the complexity of radiation protection in modern medicine: virtually all PET scanners sold today are combined PET-CT scanners where anatomical and functional images are combined. The patient has the benefit of both modalities, but the radiation protection professional must consider exposures of different natures together: radiology and nuclear medicine.

1.2.2 Therapy

Most uses of radiation for therapy are concerned with cancer treatment. However, a few nonmalignant processes are also treated using radiation: these include excessive scar formation (keloids) and unwanted growth of the conjunctiva of the eye (pterygium), which can be treated with radiation. In cancer treatment, radiotherapy aims to deliver a very high dose to the tumor while trying to minimize the dose to surrounding normal tissues. Some of the organs around the tumor may be identified as critical structures where a specific dose limit may be given to avoid unacceptable side effects. An example for the latter could be the spinal cord where a radiation dose in excess of 45 Gy (delivered in 2 Gy fractions) can result in paralysis of the patient. This problem is illustrated in Figure 1.1.

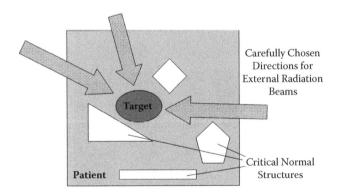

FIGURE 1.1
Illustration of the radiotherapy problem. The aim is to deliver a high radiation dose to a target inside a patient. The dose to the remainder of the patient should be held as small as possible; specifically, the dose to certain identified critical normal structures must kept below a specified dose to avoid unacceptable side effects.

There are two possible ways to deliver radiotherapy:

1. External beam radiotherapy where the radiation is directed to the cancer from outside of the patient's body. As the distance of the radiation source is typically of the order of a meter, this type of therapy is also sometimes referred to as teletherapy (*tele* is Greek for "far away").
2. Brachytherapy (*brachys* is Greek for "close by," "near") is the use of radioactive isotopes brought into close contact with the tumor to deliver the radiation. This may be through surface applicators, intracavitary applicators, or interstitial implants.

More than 90% of all radiotherapy treatments are delivered using external beam radiotherapy. However, a considerable number of radiation accidents have occurred in brachytherapy.[7] Therefore, radiation protection issues in the context of therapeutic administrations of radiation will be dealt with in two separate Chapters, 8 and 9.

1.3   Some Basic Background

1.3.1   Natural Radiation

Exposure to ionizing radiation is part of normal life. Throughout history, humans have been exposed to radiation from natural sources, such as cosmic, terrestrial, and internal radiation. This is schematically illustrated in Figure 1.2, which illustrates the relative contributions of natural and

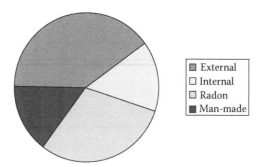

FIGURE 1.2
Typical human exposure from different sources—the contributions will vary significantly depending on lifestyle and location. The total effective dose is of the order of 3 mSv per year.

man-made radiation sources to human exposure. External radiation sources are cosmic mostly from the sun and terrestrial radiation from ground and buildings. Internal radiation (excluding radon) stems from ingested radioactive nuclides with food, such as 40-potassium. A significant contribution also comes from radon emanating from the ground. It is inhaled and as an alpha particle emitter can contribute significantly to the radiation burden. Human activity contributes at present approximately 15% to the overall radiation exposure. The human contribution is almost entirely made up by medical exposures that account for more than 90% of the exposure due to human activity, with doses from diagnostic procedures being the largest contribution.

The overall "normal" exposure in most countries is approximately 3 mSv per year. However, depending on lifestyle and where one lives and works, variations between 1.5 and 5 mSv are normal. The unit sievert (Sv) in this context is a unit measuring risk due to radiation exposure. It can be derived from the radiation dose as briefly discussed in the following and presented in more detail in Chapter 2.

1.3.2 Quantities for Radiation Protection

The **absorbed dose D** is the energy deposited per unit mass in any target material "hit" by radiation and as such measured joule per kilogram (J/kg). In the context of radiation dose, 1 J/kg is defined as 1 gray (Gy). This is the fundamental physical quantity of dose and applies to any type of radiation. The old unit for absorbed dose, which is sometimes still used in the context of radiation protection, is the rad, which equals 0.01 Gy.

The **equivalent dose H** takes into account the effect of the radiation type on tissue by using a radiation weighting factor w_R.

$$H = D \times w_R \tag{1.1}$$

A summary of weighting factors recommended by the ICRP[2] is shown in Table 2.2 in Chapter 2. The equivalent dose is measured in sievert and can be used to quantify the biological effect of dose to individual organs. The old unit for equivalent dose is the Rem, which equals 0.01 Sv.

Finally, the **effective dose E** takes into account the varying sensitivity (for stochastic effects) of different tissues to radiation using tissue weighting factors w_T according to

$$E = \Sigma_{all\,organs} (w_T\,H) = \Sigma_{all\,organs} (w_T\,w_R\,D) \tag{1.2}$$

Tissue weighting factors are summarized in Table 2.3 in Chapter 2, and the effective dose is measured in sievert. This unit is the same as the equivalent dose, and it is essential to state to what a given quantity refers, if it is not entirely clear from the context. The effective dose is used to describe the biological relevance of a radiation exposure where different organs receive a dose potentially of different magnitude from different radiation sources. As such, effective dose is essentially a quantity providing a measure of the risk associated with exposure to radiation. It is important to note that the concept of effective dose is only applicable to stochastic effects (see Chapter 3).

The notion that risk is directly proportional to radiation dose (taking the radiation type and the type of tissue hit by radiation into account) implies two important fundamental considerations for radiation protection:

1. There is no threshold for adverse effects of radiation—any dose is potentially hazardous. This is discussed in more detail in Chapter 3 in the context of the linear no-threshold model.

2. Also, natural radiation is potentially damaging, and it is impossible to distinguish effects between man-made and natural sources. The fact that humans live quite comfortably in the environment also implies that effective doses of a few microsieverts received over a long period carry risks that are manageable and difficult to distinguish from other risks that we accept.

In the context of radiation protection, sometimes the concept of **collective dose** is also useful. It is used to measure the total impact of a radiation practice or source on all the exposed persons. The collective dose is measured in man-sievert (man-Sv).

1.3.3 External and Internal Radiation Exposure

Hazards from ionizing radiation can broadly be divided in two groups:

- **External exposure:** Radiation reaches a person from outside through the skin. This is typically from a radiation source that can be turned off like an x-ray unit. The radiation from this source may cause damage in an organism while it is turned on. In the case of external exposure, nothing radioactive is left in the body. Hazards of this type may occur in radiology (see Chapter 6) or radiotherapy departments (see Chapter 8).

- **Internal exposure:** This occurs most commonly after the incorporation (e.g., breathing in, consuming with food, absorbing through the skin) of radioactive isotopes. The radioactivity remains in the organism until the isotope has decayed (physical half-life) or until it is excreted (e.g., in urine or during exhalation). These hazards may be present in nuclear medicine departments (see Chapter 7). Internal exposure may also be of concern in research laboratories in a hospital environment. These are not specifically covered in the present book.

In brachytherapy typically only sealed sources are used. Therefore, the risk of internal exposure is small. The use of open radioactive sources for therapy is covered in Chapter 7 as part of radiation protection in nuclear medicine.

1.3.4 The Radiation Warning Sign

Ionizing radiation cannot be seen, heard, smelled, or otherwise sensed by humans. However, as radiation is considered a hazard, it is important to make people aware of its presence. The internationally recognized symbol indicating radiation hazards is the black trefoil on a yellow background. This is shown in Figure 1.3. Recently a new supplementary warning sign has been launched by the IAEA. It is designed to be used only as indoor housing and close to high-activity radioactive sources; it is meant to be more self-explanatory and easier to understand immediately (http://www.iaea.org/NewsCenter/News/2007/radiationsymbol.html, last accessed May 2007). It is also shown in Figure 1.3.

1.4 The Framework of the 1990 Recommendations of the ICRP

1.4.1 The ICRP

Most radiation protection legislation in the world is based on the recommendations of the International Commission on Radiological Protection (ICRP). The commission was formed in 1928 with the brief to develop guidelines and units for radiation protection. As the ICRP Web page puts it:

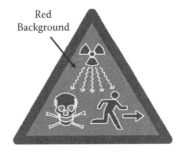

FIGURE 1.3
Radiation warning signs. *Left*: The international radiation symbol (black symbol on yellow background). *Right*: Supplementary warning sign recently introduced by the IAEA (black on red background).

"The International Commission on Radiological Protection, ICRP, is an independent Registered Charity, established to advance for the public benefit the science of radiological protection, in particular by providing recommendations and guidance on all aspects of protection against ionising radiation." (www.icrp.org)

It also has the task to publish data that can further the field of radiation protection. A good historical overview has recently been given by R. Clarke and J. Valentin,[8] the long-time chairman and secretary of the ICRP.

The ICRP is funded by voluntary contributions and independent of governmental influences. It defines the philosophy underlying radiation safety and recommends limit values; it is the role of national radiation protection organizations and local authorities to see these through into a regulatory framework. As the web page of the ICRP states, the ICRP

is composed of a Main Commission and five standing Committees: on Radiation effects, on Doses from radiation exposure, on Protection in medicine, and on the Application of ICRP recommendations, and on Protection on the environment, all served by a small Scientific Secretariat. The Main Commission consists of twelve members and a Chairman (currently Dr L-E Holm, Sweden). Like other scientific academies, the Commission elects its own members, under rules that are subject to the approval of ISR (International Society of Radiology). Renewal is assured in that 3 to 5 members must be changed every fourth year. Committees typically comprise 15–20 members. Biologists and medical doctors dominate the current membership; physicists are also well represented. (www.icrp.org)

Recommendations of the ICRP form the basis of radiation protection regulations in most countries of the world. As such, the publication of Report 60, *1990 Recommendations of the International Commission on Radiological Protection*, published in 1991, found its way over the years into legislation in many nations. An important publication in this context is the *Basic Safety Standards* (BSS) published in 1996 by the International Atomic Energy Agency[9] and adopted, among others, by the World Health Organization (WHO) and the International Labor Organization (ILO). Appendix II in the BSS is dedicated to medical exposures and as such most closely linked to the present book. The *Basic Safety Standards* are an internationally binding document that is based on ICRP Report 60 (1991).

ICRP Report 60 defines the philosophy underlying radiation safety, gives extensive rationale for the scientific basis, and provides practical recommendations, for example, in terms of limit values and dose concepts. In March 2007, the commission approved a new set of recommendations (which were not published at the time of this writing). However, a press release and versions circulated for comments indicate that the basic concepts and framework of the 1990 recommendations[2] are likely to still be valid. New scientific evidence is incorporated in the document as well as a novel emphasis on protection of the environment added (http://www.icrp.org).

1.4.2   Types of Radiation Exposure

According to the ICRP, it is useful to distinguish among three classes of radiation exposure that need to be managed differently: occupational, medical, and public. These categories were defined by the ICRP[2] and subsequently included in the IAEA *Basic Safety Standards:*[9]

- **Occupational exposure** is incurred at work by an individual knowingly working with ionizing radiation. The exposure is a result of the nature of the work, and the occupationally exposed persons are typically monitored for exposure to ionizing radiation (see Chapter 5) and have regular training in safe practices to do with ionizing radiation.
- **Public exposure** is the exposure of any member of the general public. Public exposures cover all exposures arising from a particular activity involving the use of radiation.
- **Medical exposure** is the exposure of patients as part of diagnostic procedures or as part of their therapy. Medical exposures also include volunteers in medical research. Only in this case is it possible to apply meaningful dose constraints; in the case of patients the prescription provides justification, and it is always assumed that the potential benefits outweigh the risks for the individual patient.

The first two categories are easy to appreciate, and the three general principles of radiation protection discussed above apply directly. For medical exposures a dose limit value as such should not apply. The exposure always has to be seen in the context of the benefits to the patient, and the prescription provides justification for the exposure.[10] However, medical exposures also apply to persons providing support for others who undergo a medical exposure (e.g., parents supporting children) and volunteers in medical research. In both these cases, dose constraint values must be considered, and national legislation/regulation usually provides for this.

1.4.3   Principles of Radiation Protection

Report 60 of the ICRP[2] defines three underlying principles of radiation protection: justification of practice, optimization of protection, and individual dose limits.

1.4.3.1 Justification of Practice

No practice involving exposures to radiation should be adopted unless it produces sufficient benefit to the exposed individuals or to society to offset the radiation detriment it causes. This thinking is based on the assumption (discussed above and in more detail in Chapter 3) that stochastic radiation effects may occur at any dose level, and even very small radiation doses are potentially harmful.

For example, in medical exposures, such as occurs with patients undergoing radiotherapy, the justification is given by the prescription of the radiation oncologist. As such, the prescription is a very important document from a radiation protection point of view.

1.4.3.2 Optimization of Protection

This principle requires that all exposures be optimized to maximize the benefit and minimize the risks. This is the most important principle for radiation protection practice; here one can make a real difference when handling exposures well.

For exposures, other than when therapeutic benefit is intended, optimization of protection can be described by the so-called ALARA principle (as low as reasonably achievable). This means radiation exposure should be limited as much as possible, keeping in mind the risk-benefit relation of radiation and its applications. For example, it is unreasonable to refuse a radiograph after a bone fracture just because statistically this may shorten your life expectancy by 1 day. The benefits of the radiograph with its diagnostic value far outweigh the tiny risk associated with the radiation exposure. The optimization of protection does not apply as intuitively to the dose given to the tumor in radiotherapy, where the dose delivery is the objective of the treatment. However, even in this case, optimization of protection will require the operator to minimize the dose to normal structures while retaining compatibility with the therapeutic aim. In all circumstances—diagnostic and therapeutic—optimization includes the concept of minimization of the risk of accidents.

It is important to note that ALARA requires consideration of the societal and economic contexts. As such, it cannot be applied to all situations in the same way, and professional judgment may be required to achieve an outcome that is optimal for a particular situation.

1.4.3.3 Individual Dose and Risk Limits

In addition to the two principles discussed above, the ICRP recommends dose limits for persons exposed to ionizing radiation. These limits are meant to restrict the risk due to exposure to the same order of magnitude as other risks we generally accept (such as driving a car) or are expected to take as part of employment (driving a taxi). The limit values recommended by the ICRP in Report 60 are 1 mSv/year for the general public and 100 mSv/5years for occupationally exposed persons, with the effective dose in any single year not to exceed 50 mSv. In addition to this, the commission provides specific annual limits for equivalent doses to the lens of the eye (15 mSv public, 150 mSv occupational), the skin (50 mSv public, 500 mSv occupational), and the hands and feet (only defined for occupationally exposed persons: 500 mSv).[2]

It is important to note that these dose limits apply in addition to the dose received by any person from natural radiation sources.

1.5 Organizing Radiation Protection in a Medical Environment

The concept of a radiation protection program includes a number of measures to reduce the likelihood of adverse effects due to the use of ionizing radiation. A useful publication for the implementation of such a program in the context of radiotherapy is Technical Document (TECDOC) 1040 of the IAEA.[11]

A radiation protection program is considering a facility as a whole. Figure 1.4 illustrates the relationships between all parties involved in a radiation safety program. The regulatory authority will issue a license for the use of ionizing radiation for a particular purpose, such as provision of radiotherapy services. The licensee is the person (in a legal sense) who is responsible for radiation safety. This implies that it must be of interest to managers to provide sufficient resources to radiation protection. Within the institution, radiation protection affects many parties, as indicated in the figure. It

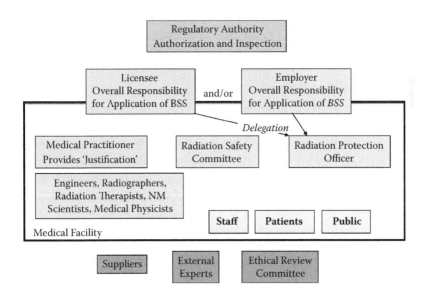

FIGURE 1.4
Participants in a radiation protection program as discussed in the IAEA Basic Safety Standard 1996.[9]

is important to consider also visitors, contractors, and other outside persons. They are generally harder to inform about radiation hazards, and it will be difficult to provide any meaningful training and monitoring.

1.5.1   A Radiation Protection Program

Typical elements of a radiation protection program are the following:

- Assignment of responsibilities: The registrant/licensee is the legal person responsible for radiation protection. The general terminology is that a radiation practice of small risk can be registered while a more complex practice with higher risks must be licensed.[9] In practice, though, some responsibilities can be delegated, for example, to the radiation protection officer (RPO).

- Radiation protection officer: The RPO is a person technically competent to provide advice on and oversight of the local radiation safety program. He or she is also often referred to as the responsible person. This person is often (but not necessarily) a physicist. He or she is a crucial component of the program and should be given the resources and authority necessary. The role would typically include responsibility for designation of controlled and supervised areas, responsibility for ensuring preparation of local rules, the training of new staff in safe radiation work practices, liaison with the regulatory authority on radiation protection matters, supervision of the personnel monitoring program, and maintenance of records, especially worker radiation histories. In addition to this, he or she would perform routine surveillance of radiation areas and respond to radiation incidents and accidents. Many of these issues are discussed in more detail in Chapter 5.

- Radiation safety committee (RSC): Typical roles of the RSC are to oversee the institutional radiation safety program and advise and review local rules relevant for radiation protection. Members may require special training.

- Designation of radiation areas: See Chapter 5. This includes appropriate warning signs and radiation indicators such as the ones shown in Figures 1.2 and 1.5.

- Local rules: These are intended to provide adequate levels of protection and safety through the establishment of common work procedures and other systems to be followed by all workers in the area. They should be set down in writing and include all information required for work in the area and be made known to all workers.

- Education and training: All staff in radiation areas must have appropriate education to perform their duties. Staff who initiate radiation must receive training as well as persons who could be subject to irra-

FIGURE 1.5
Radiation beam-on lights in a radiotherapy bunker. The sign indicates not only that radiation is on but also if it is imminent or if it is safe to enter the room. The sign is large enough and easy in design to be immediately understood by members of the public. It may also be coupled with an audible alarm.

diation. Technology is fast developing. It is therefore essential for all staff to have regular updates on radiation protection aspects.

- Planning for accidents and emergencies: This planning needs to be done in cooperation with other departments and groups. For example, fire and emergency services should have a plan of the facility and be aware of any potential radiation hazards.

- Health surveillance and monitoring: This is discussed in more detail in Chapter 5.

- Review and audit: This is discussed in more detail in Chapter 5.

- System of recording and reporting: In each radiation facility there should be a system instituted where all relevant information relating to radiation work is recorded, documented, and, when necessary, reported to management and the regulatory authority as required. This is a key factor in control of exposures and maintenance of a safe working environment and may depend on national regulations.

- Radiation protection manual or management plan: This is discussed below.

1.5.2   Radiation Safety Manual

Most of the procedures concerning radiation protection can be summarized in a radiation protection manual now also referred to as a radiation management plan. The manual is ideally a local reference book and made available to (and possibly mandatory reading for) all radiation workers in local language if appropriate. In general, it can be expected that the radiation safety manual includes sections on:

- Basics of radiation safety
- Sources, risks, and effects of radiation
- Local radiation safety organizations
- National/state regulations
- Personnel monitoring
- Emergency procedures
- Local rules in radiation-user departments (would typically be the same areas as covered in Chapters 6 to 9 in the present book)
- Death procedures (patients containing radioactive materials)
- Radiation and pregnancy (for radiation workers, but also including information relating to pregnant patients)
- Incident/accident procedures
- Research and radiation

This list is also a good indication of issues that need to be considered for radiation protection in medicine. Many of these aspects apply to all radiation areas in a hospital. They will be addressed in more detail in relation to the different activities in medicine discussed more fully in subsequent chapters of this book.

References

1. Walsh, D., Deep tissue traumatism from roentgen ray exposure, *Br. Med. J.*, July 31, 272–73, 1897.
2. International Commission on Radiological Protection, *1990 Recommendations of the International Commission on Radiological Protection*, ICRP Report 60, Oxford, UK: Pergamon Press, 1991.
3. Clarke, R., 21st century challenges in radiation protection and shielding: draft 2005 recommendations of ICRP, *Radiat. Prot. Dosimetry*, 115, 10–15, 2005.
4. Clarke, R. H., Draft recommendations from ICRP at the start of the 21st century, *Health Phys.*, 87, 306–11, 2004.

5. Kron, T., Wilhelm Conrad Röntgen, *Australas. Phys. Eng. Sci. Med.*, 18, 121–23, 1995.
6. Macklis, R. M., Portrait of science. Scientist, technologist, proto-feminist, superstar, *Science*, 295, 1647–48, 2002.
7. International Commission on Radiological Protection, *Prevention of Accidental Exposures to Patients Undergoing Radiation Therapy*, ICRP Report 86, Oxford: Pergamon Press, 2001.
8. Clarke, R., and Valentin, J., A history of the international commission on radiological protection, *Health Phys.*, 88, 717–32, 2005.
9. International Atomic Energy Agency, *International Basic Safety Standards for Protection against Ionizing Radiation and for the Safety of Radiation Sources*, IAEA, Vienna, 1996.
10. International Commission on Radiological Protection, *Radiological Protection and Safety in Medicine*, ICRP Report 73, Oxford: Pergamon Press, 1996.
11. International Atomic Energy Agency, *Design and Implementation of a Radiotherapy Programme: Clinical, Medical Physics, Radiation Protection and Safety Aspects*, TECDOC 1040, IAEA, Vienna, 1998.
12. ICRP, ICRP Publication Number 103: Recommendations of the ICRP. Elsevier, 2007.

2

Fundamentals of Radiation Physics

Jamie V. Trapp and Peter Johnston

CONTENTS

The aim of this chapter is to provide an overview of radiation physics and the units used in radiation protection. It is intended to provide the reader with basic information to form the understanding required to take full advantage of the later chapters. We start with an overview of the origins of radiation and move on to the interactions of radiation with matter. These concepts will then be extended in Chapter 3 to describe how radiation affects the human body.

2.1   Origins of Radiation

2.1.1   Nuclear Structure

The logical starting point when discussing radiation is its creation, or emission from an atom. The simplest atomic model to use for this purpose is the Bohr model, which consists of a nucleus of protons with positive electric charge and neutrons with no net electrical charge. The nucleus is surrounded by negatively charged electrons that orbit at given distances related to the energy of the electron, as shown in Figure 2.1. There can be a number of electrons with the same orbital radius, and this radius is said to be an *electron shell*. The electron shells are labeled, from innermost outwards, K, L, M, N, etc. Electrons in the K shell are bound the most tightly, requiring the most energy to remove from the atom, and the binding energy subsequently decreases from inner to outer shells, so outer shell electrons are most loosely bound. The Bohr model is generally considered a layman's view of the structure of an atom: quantum mechanics show that the structure of an atom does not truly follow the Bohr model. Despite this fact, the Bohr model of the atom is suitable for fundamental descriptions of the origins of radiation, including

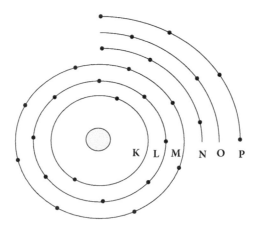

FIGURE 2.1
Bohr model of an atom, consisting of a nucleus containing protons and neutrons, surrounded by a series of shells containing electrons.

Mass Number

$$_Z^A X$$

Atomic Number

Chemical Symbol

FIGURE 2.2
The conventional written form for a radionuclide.

diagrammatical representation and simple calculations, and so will be used throughout this chapter.

Chemical elements are often described in terms of their name, atomic number, Z, and mass number, A, and usually written in the form shown in Figure 2.2. The atomic number is the total number of protons in the nucleus and the mass number is the total number of nucleons (protons and neutrons).

The element is defined by the atomic number, as it is the charge of the nucleus that determines the element, whereas variations in the number of neutrons in a nucleus will not change the element, but will produce different *isotopes* of the same element. For example, gold has only one stable isotope, but thirty-one radioactive isotopes have been observed. The number of neutrons in observed gold isotopes varies from 94 to 125, but each isotope of gold always has 79 protons. If ^{197}Au gains an extra neutron it will become ^{198}Au, which is a different isotope of gold; however, if ^{197}Au gains an extra proton it will become ^{198}Hg, which is no longer an isotope of gold but an isotope of mercury. When the number or type of nucleons spontaneously changes within a nucleus, whether the change is to a new element or, very rarely, a different isotope of the original element, the nucleus is said to have undergone radioactive decay.

For a "neutral" atom, the number of electrons filling the electron shells matches the number of protons in the nucleus, and hence the atom as a whole has no net electric charge. If an electron is either gained or lost without any change to the number of protons, the net electric charge of the atom will no longer be zero, and thus the atom becomes *ionized* and is called an *ion*. Changes in the structure of either the electron shell or the nucleus often result in the emission of energy by a radiative process.

2.1.2 Decay Rates and Half-Life

If a material contains a number of radioactive nuclei, the average decay rate, A, known as the activity, is always proportional to the number of nuclei present, N:

$$A = \lambda \, N \qquad (2.1)$$

where λ is termed the decay constant. For example, of a 1 mg sample of ^{238}U atoms, containing 2.5×10^{18} atoms, about twelve nuclei, will decay over the following second. If the sample contained half of the number of ^{238}U atoms, only about six nuclei would decay in the following second. Decay is a random process, which means that there is no way to predict which of the nuclei in a sample are going to decay, and while the decay rate itself undergoes minor fluctuations, the average decay rate is related to the number of nuclei present.

Because the decay rate is proportional to the number of atoms of an iso-tope present, and because by definition the decay itself causes the number of nuclei to continually change, the decay rate will also continuously change. For example, take a radionuclide of which 20% of nuclei decay over a certain period of time. If the initial number of nuclei present is one hundred, at the end of the time period twenty will have decayed, leaving eighty nuclei. Over the next time period, 20% of the remaining eighty nuclei will decay (not 20% of the original one hundred), which will now leave sixty-four of the original nuclei present. For the third time period 20% of the remaining sixty-four nuclei will decay, and so on. This means that, over time, both the number of original nuclei and the decay rate will not reduce at a constant rate, but will reduce exponentially:

$$N(t) = N_0 e^{-\lambda t} = N_0 2^{\frac{-t}{T_{1/2}}}, \qquad R(t) = R_0 e^{-\lambda t} = R_0 2^{\frac{-t}{T_{1/2}}} \qquad (2.2)$$

where N_0 and R_0 are the initial number of nuclei and the initial decay rate, and $T_{1/2}$ is the *half-life*, which is the time taken for the number of nuclei of a particular radioisotope to decay to half of its initial number. Known half-lives vary from femtoseconds (10^{-15} s) to more than 10^{19} years. The most radioac-tive nuclei have very short half-lives.

Example 2.1
Q: Iodine 125 is widely used in nuclear medicine and has a half-life of 60 days. Calculate the decay constant, the decay rate of a 4 MBq source after 120 days, and how long will it take for an ^{125}I source to decay to 4% of its initial decay.

A: The decay constant is first calculated:

$$\lambda = \frac{\ln(0.5)}{60 \times 24 \times 3600} = 1.34 \times 10^{-7} \, s^{-1}$$

The decay rate of a 4 MBq source after 120 days:

$$R = 4 \times e^{-\lambda t} = 4 \times e^{1.34 \times 10^{-7} \times 120 \times 24 \times 360} = 625 kBq$$

How long will it take for an ^{125}I source to decay to 4% of its initial decay?

$$\frac{A}{A_0} = e^{-\lambda t}$$

$$0.04 = 4 \times e^{-1.34 \times 10^{-7} t}$$

$$t - 278 \ days$$

2.1.3 Electromagnetic Radiation

Electromagnetic radiation covers an extremely broad range of energies, from very long wavelengths (extra-low-frequency radiation) to radio waves, infrared light, visible light, ultraviolet light, x-rays, and gamma rays. A quantum, or particle, of electromagnetic radiation is termed a photon. This book deals with radiation protection from radiation with enough energy to ionize an atom, that is, ionizing radiation.

Ionizing electromagnetic radiation is often encountered in the form of x-rays or gamma rays. Essentially, there is no difference between an x-ray photon and a gamma photon once they are in existence, and the name refers only to their origin. X-rays originate from changes in the arrangement of electrons; that is, an electron "jumps" between electron shells of different energy—due to conservation of energy a photon is emitted carrying away the excess energy. Gamma rays originate from changes within the nucleus, that is, reconfiguration of the nucleons to achieve a lower-energy configuration. Because nuclear changes normally involve greater energy transitions than electronic transitions, gamma rays are frequently, but not always, of greater energy than x-rays.

2.1.4 Alpha Radiation

Alpha radiation is the emission of two protons and two neutrons in the form of a ^{4}He nucleus from the parent atom. Alpha decay occurs with discrete energies, rather than as a continuous spectrum, and this energy is in the form of kinetic energy of the alpha particle and the recoiling daughter nucleus. Conservation of momentum determines how much energy the alpha particle and the daughter will carry, with the alpha particle carrying most of the decay energy.

2.1.5 Beta Radiation

Beta radiation, or beta decay, is the emission of either an electron or a positron from the nucleus of an atom (a positron is a positively charged electron, i.e., the antiparticle of the electron) as shown in equation (2.3a) and (2.3b). A variation is the process of *electron capture*, where the nucleus captures an inner shell electron as shown in equation (2.3c) and is usually accompanied by the emission of x-rays as the vacancy in the inner shell is filled, and if the nucleus de-excites after electron capture, then one or more gamma photons will also be emitted.

$$\frac{A}{Z}X \to \frac{A}{Z+1}Y + \frac{0}{-1}e + \bar{\upsilon} \tag{2.3a}$$

$$\frac{A}{Z}X \to \frac{A}{Z+1}Y + \frac{0}{-1}e + \upsilon \tag{2.3b}$$

$$\frac{A}{Z}X + \frac{0}{-1}e \to \frac{A}{Z+1}Y + \upsilon \tag{2.3c}$$

The extra terms, υ and $\bar{\upsilon}$ seen in equation (2.3a) and (2.3b), relate to the emission of an elementary particle called a neutrino, which always accompanies beta decay and is necessary to conserve energy and momentum. Unlike alpha decay, which is always of a discrete energy, beta particles are emitted from an atom with a continuous spectrum of energies. This spectrum of energy is explained by the fact that there are three exiting particles: the electron, neutrino, and recoiling nucleus, each of which carries away part of the energy from the nuclear decay.

2.2   Radiation Interactions and Energy Deposition in Matter

2.2.1   Photons

If a monoenergetic, parallel beam of photons is incident on a material, the number of these photons in the beam (the intensity) will exponentially decrease with the distance that the beam has traveled through the material and is termed Lambert's law:

$$I = I_0 e^{-\mu t} \tag{2.4}$$

where I and I_0 are the transmitted and initial intensity of the beam, respectively, μ is the *linear attenuation* coefficient, and t is the thickness of the material through which the beam has passed. The linear attenuation coefficient is dependent on both the energy of the incident radiation and the density of the absorbing material. Therefore, the mass attenuation coefficient (μ/ρ, where ρ is density) is often more convenient to use as it is independent of the physical form or state of the material. It is important to note that only the number of photons decreases as the photon beam travels through the material—the energy of the original monoenergetic photon beam does not change. Furthermore, if a photon beam passes through a slab of material, each individual photon has a probability of interacting within the material; hence each photon will either pass through the material undisturbed or interact.

The main processes through which ionizing radiation photons interact with matter leading to energy deposition in matter are the photoelectric effect, Compton scattering, and pair production. Although other interactions between photons and matter occur, they are not discussed here as they mainly apply to photons with energies outside the range of interest to medicine.

The photoelectric effect occurs when an incident photon interacts with an atomic electron, to which it transfers all of its energy (Figure 2.3). The electron is then ejected from the atom and will have a kinetic energy that is the

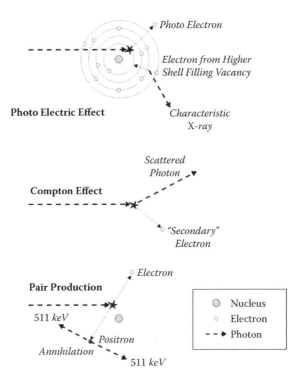

FIGURE 2.3
Diagrammatic representation of common photon interactions with matter. The photoelectric effect (*top*) occurs when an incident photon transfers all of its energy to an atomic electron. Compton scattering (*middle*) occurs when an incident photon transfers part of its energy to an electron, and the remainder energy is carried away as a scattered photon. In pair production (*bottom*) an incident photon interacts with an atomic nucleus and disappears to create an electron-positron pair. The positron annihilates with a nearby electron and results in two annihilation photons traveling in opposite directions.

difference between the energy of the incident photon and the energy that bound the electron to the atom:

$$E_e = E_{photon} - E_b \tag{2.5}$$

where E_e is the energy of the photoelectron, E_{photon} is the energy of the incident photon, and E_b is the binding energy of the electron prior to ejection from the atom. The photoelectric effect is most probable for incident photons with energies less than 100 keV, and it occurs preferentially with the most tightly bound electrons. In addition, there is a preference for photoelectric absorption to occur with heavier nuclei than lighter, with the probability increasing rapidly with atomic number (approximately Z^4).

The Compton effect, or Compton scattering, also involves interaction between an incident photon and an atomic electron resulting in the ejection of an electron from an atom; however, the incident photon transfers only part of its energy to the ejected electron, and the remainder of the energy is carried away as a secondary scattered photon (see Figure 2.3). In medical applications these secondary scattered photons contribute to the radiation dose in patients, and in diagnostic radiology collimating grids are often used to reduce the loss of contrast in images arising from the undesirable effects of scatter.

The energy of a scattered photon is a function of the angle between the directions of travel of the incident and scattered photons, and is given by

$$E'_\gamma = \frac{E_\gamma}{\sqrt{1 + \left(E_\gamma / mc^2\right)\left(1 - \cos\right)}} \tag{2.6}$$

where E'_γ is the energy of the scattered photon, E_γ is the energy of the incident photon, θ is scattering angle, and m is the mass of an electron. The probability of a scattered photon traveling in a certain direction is given by the *Klein-Nishina* formula for the differential scattering cross section per electron $d\sigma_e/d\Omega$:

$$\frac{d\sigma_e}{d\Omega} = r_0^2 \left[\frac{1}{1 + \alpha\left(1 - \cos\theta\right)}\right]^3 \left[\frac{1 + \cos\theta}{2}\right]\left[1 + \frac{\alpha^2\left(1 - \cos\theta\right)^2}{\left(1 - \cos^2\theta\right)\left[1 + \alpha\left(1 - \cos\theta\right)\right]}\right] \tag{2.7}$$

where α is the photon energy in units of electron rest energy (E_γ/mc^2) and r_0 is a parameter called the classical electron radius (2.818 fm). High-energy (>1 MeV) photons are most likely to be mainly scattered in the forward direction, while for lower energies the scattered photons tend to scatter in all directions.

The third major interaction mechanism of ionizing photons is pair production. In this case an incident photon interacts with the nucleus of an atom and disappears completely, with its energy transferred to the creation of an

TABLE 2.1

Photon Interaction with Matter for the Energy Range of Gamma and X-radiation

PRIVATE Interaction Type	Dependence on Atomic Number	Dependence on Photon Energy	Secondary Particles
Classical scattering	Z2.5/A	E^{-2}	
Photo effect	Z4/A	E^{-3}	Electrons, characteristic x-rays, Auger electrons
Compton effect	Z/A	$E^{-1/2}$	Electrons, scattered photons
Pair production	Z2/A	E > 1022 keVlog E	Electrons, positrons, annihilation radiation
Nuclear photo effect	Depends on material	E > threshold for particular material	Neutrons, protons, fission, etc.

Note: Z = atomic number, A = atomic mass; except for hydrogen one can use as a first approximation A = 2Z. The atomic number and energy dependence are approximations only and vary with each other.

electron-positron pair. Although electrons and positrons individually have electric charge, while the incident photon does not, the electron-positron system as a whole has no net charge and so conservation of electric charge is maintained. Both newly created particles equally share the energy of the incident photon and are free particles (not bound to the nucleus) and are therefore free to interact with their surroundings. The electron and positron lose energy through interactions with their surroundings. After the particles thermalize, the electron will normally be captured by a nearby atom and the positron will tend to annihilate with a nearby electron. The annihilation process results in the creation of two annihilation quanta (photons) that travel in opposite* directions, as represented in Figure 2.3. The rest mass of an electron or positron is 0.511 MeV; therefore, for pair production to occur the energy of the incident photon must be equal to or greater than 1.022 MeV.

The above interactions are summarized in Table 2.1. From the discussion, it can be concluded that human exposure to ionizing photons can be reduced by designing a radiation shield that is not only thick (due to the exponential decrease in photon penetration with distance) but also of a material with high atomic number.

Example 2.2

Q: The 1.17 and 1.33 MeV gamma radiation from ^{60}Co has been used for many decades for treatment of cancerous tumors. Using the mass atten-

---

* The small net momentum of the electron-positron annihilating pair can cause the directions of the quanta to be not exactly opposite, but this is of no practical importance in medical physics.

uation coefficient of 6.265×10^{-2} cm^2/g for 1.25 MeV photons in soft tissue and density of 1.060 g/cm^3, estimate what percentage of a narrow beam of incident photons will be absorbed in a 20 cm thick portion of soft tissue in a patient.

A:

$$I = I_0 e^{-\frac{\mu}{\rho}\rho t}$$

$$I = 100(\%)e^{-\frac{6.265\times10^{-2}}{1.060}\times1.060\times20}$$

$$I = 26\%$$

With the intensity of 26% of incident photons exiting the tissue, 74% are absorbed within the tissue. Note that this example applies to narrow-beam geometry; in broad-beam geometry other factors such as scattered radiation will contribute to radiation dose in the patient.

2.2.2   Alpha Particles

When charged particles interact with matter the energy loss process is somewhat different from that of photons. An individual photon will either not interact at all or interact once and be removed from the beam. When a large charged particle such as an alpha particle transits a material it undergoes many interactions in a process that appears, on a large scale, to be a continuous energy loss until the kinetic energy is finally lost; macroscopically this appears smooth, as is described using the *continuous slowing down approximation* (CSDA). The dominant mechanism for this energy loss is Coulomb scattering with atomic electrons (it is approximately 10^{12-15} times more likely that an alpha particle will react with an electron than with a nucleus).

In a head-on collision between an alpha particle and an electron the maximum kinetic energy transfer is given by the following equation:

$$T = T\left(\frac{4m_e}{M_a}\right) \tag{2.8}$$

where T represents kinetic energy of the alpha particle, m_e is the mass of an electron, and M_a is the mass of the alpha particle. The ratio of the mass of an electron to that of an alpha particle is 1:1,836, which means that a megaelectronvolt alpha particle will undergo many thousands of collisions before losing all of its energy. As the Coulomb force has an infinite range, the alpha particle will usually be interacting with many electrons at the same time in a continuous energy loss process. Because the energy loss for each interaction is small in comparison with the initial energy, there will be little deflection of the particle along its path, and so it will travel in an almost straight line.

The distance that an alpha particle travels before losing all of its energy is called the *range*. The range of alpha particles is a function of not only the initial energy of the particle, but also the material through which it is traveling. Given a particular initial energy and material, the range of an alpha particle can be calculated; however, a statistical variation occurs for each individual particle due to different particle histories; that is, individual particles travel slightly different paths with slight variations in the distance between interactions. The mean range is defined as the distance at which the intensity of a beam of alpha particles is halved (Figure 2.4). Although alpha particles undergo many thousands of interactions before losing all of their energy, the distance between the electrons with which the alpha particle is interacting is very small, and hence the range is quite small. For example, the range of a 5 MeV alpha particle in air is approximately 35 mm while in tissue it is approximately 0.04 mm; hence, the most practical shield to reduce human exposure to alpha particles is often the skin. Health hazards from alpha particles are therefore normally only through inhalation or ingestion.

2.2.3   Beta Particles

Like alpha particles, electrons interact with matter through the Coulomb force. However, due to the much smaller mass of the electron, the paths followed by electrons are complex. Where alpha particles traverse a material in an almost straight path due to their mass, electrons undergo large angular deflections due to the fact that they are interacting with atomic electrons

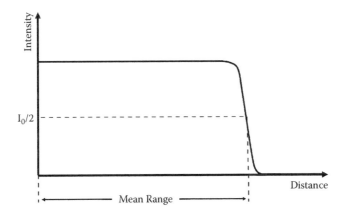

FIGURE 2.4
Mean range of alpha particles. Although the range of an alpha particle can be predicted, statistical variations will occur between individual particles due to different particle histories.

with similar mass, even considering relativistic effects. The result of this is that electrons travel in quite erratic paths during the energy loss process.

The implications arising from the erratic path of electrons in matter are, first, that the range of electrons is defined as the linear distance that the electron penetrated into the material, and not the actual path length through which the electron traveled in its erratic journey. Despite this, the range of electrons is greater than that of alpha particles of similar energy due to the fact that electrons have a much smaller mass and therefore a much higher velocity for a given energy. In tissue the range of electrons tends to be of the order of a few millimeters, compared to the fractions of millimeters of alpha particles.

Second, the large angular deflections of electrons in matter result in the emission of electromagnetic radiation, which occurs whenever charged particles are accelerated. When applied to electrons, this phenomenon is called *bremsstrahlung* (braking radiation). This is important in medical applications, as x-ray beams used in diagnostic radiology and radiation therapy are actually bremsstrahlung beams produced by accelerating electrons onto a material of high atomic number such as Tungsten. Once the bremsstrahlung photons are produced they interact with matter through the same processes as other electromagnetic radiation, described earlier.

2.2.4   Neutrons

Neutrons are nucleons that are slightly heavier than protons; however, they do not have a net nuclear charge, and therefore do not interact with matter through the Coulomb force. This means that they lose energy through elastic collisions with other nuclei.

Neutrons are classified according to their kinetic energy as fast neutrons (over 0.1 MeV), slow neutrons (up to 0.5 eV), or thermal neutrons (0.1 eV or less). Fast neutrons can undergo elastic or inelastic collisions with nuclei. Elastic collisions can be likened to a billiard ball collision where the atomic nucleus undergoes recoil. The maximum energy transfer with such a collision occurs with nuclei of similar mass to a neutron, such as hydrogen, which has only a single proton as the nucleus. The implications of this are that shielding of neutrons is best achieved with hydrogen-rich materials such as paraffin (as opposed to high-atomic-number materials required for photons and electrons) and that neutrons can be a health hazard to humans due to the large water content in tissue. Inelastic collisions occur when the neutron collides with a nucleus, with some of the energy being given to excitation of the nucleus, often followed by prompt de-excitation with the excess energy radiated as a photon.

Elastic scatter is possible for thermal neutrons; however, it is more probable that neutron capture by the nucleus will occur. In this reaction the neutron is absorbed into the nucleus and often results in a radioactive species that may emit either a photon or particle as it de-excites.

2.3   Units in Radiation Protection

The previous section introduced the fundamentals of radiation interactions with matter. In this section these concepts are progressed to quantify exposure to radiation and the subsequent deposition of radiation energy in matter.

2.3.1   Absorbed Dose

From a human health perspective, the most important result of radiation interactions with matter is the deposition of energy in tissue. The energy transferred to tissue can be in the form of heat, ionization of individual atoms, or changes in the properties of the molecules contained within the tissue. Before discussing the biological effects of this energy transfer in Chapter 3, it must first be quantified.

Absorbed radiation dose describes the amount of energy deposited per unit mass. The SI unit for absorbed dose is the Gray (Gy), expressed in terms of joules per kilogram:

$$1 \text{ Gy} = 1 \text{ J/kg} \tag{2.9}$$

Prior to the introduction of the Gray, the generally accepted unit of absorbed radiation dose was the rad, expressed as ergs per gram:

$$1 \text{ rad} = 100 \text{ erg/g} \tag{2.10}$$

Conveniently, conversion between rad and Gy is simple:

$$100 \text{ rad} = 1 \text{ Gy} \tag{2.11}$$

or

$$1 \text{ rad} = 1 \text{ cGy} \tag{2.12}$$

Example 2.3

Q: A beam of ionizing radiation is incident on a 1.5 L container of water and causes a temperature rise of 0.05°C throughout the water. Calculate the absorbed dose in the water.

A: The first step is to calculate the energy absorbed in the water. Given that the specific heat of water is 4186 J/kg·°C and the mass of 1.5 L of water is 1.5 kg, the absorbed energy is

$$(1.50 \text{ kg})(4186 \text{ J/kg} \cdot °C)(0.05 \text{ °C}) = 313.95 \text{ J}$$

When the absorbed energy is known, the absorbed dose can be calculated according to equation (2.9):

$$313.95 \text{ J} / 1.50 \text{ kg} = 209.28 \text{ Gy}$$

2.3.2   Exposure Units

To receive an absorbed radiation dose, one must first actually be exposed to radiation, and so a quantity is defined for radiation exposure. This quantity is calculated in terms of the amount of ionization per mass of air by x-rays or gamma rays up to energies of around 3 MeV. The SI unit for this quantity is called the exposure unit, or X unit:

$$1 \text{ X} = 1 \text{ C/kg (air)} = 33.97 \text{ J/kg (air)} \tag{2.13}$$

Prior to the introduction of SI units, the unit for exposure was the roentgen (R):

$$1 \text{ R} = 2.58 \times 10^{-4} \text{ C/kg (air)} \tag{2.14}$$

The roentgen is based on 1 electrostatic unit of ionization (3.36×10^{-10} C) per 1 cm^3 of air at 0°C and 1 atmosphere (0.001293 g).

2.3.3   Equivalent and Effective Dose

In many cases, simply measuring the absorbed radiation dose provides insufficient information for predicting the probability of stochastic effects within an individual. This is because the biological effect of a particular radiation dose varies depending upon both the type and energy of radiation and the tissue in which the radiation is absorbed. To cope with this, the International Commission on Radiological Protection (ICRP) has defined the terms *equivalent dose* and *effective dose*.

2.3.3.1 Equivalent Dose

Equivalent dose is the term used to describe the differing effects of different types and energies of radiation on the same tissue. For example, most tissue is much more sensitive to radiation in the form of heavy charged particles such as alpha radiation than it is to electromagnetic radiation such as gamma radiation or x-rays. Therefore, an absorbed dose of 1 cGy from alpha radiation would be more damaging to tissue than an absorbed dose of 1 cGy from x-rays, and so the equivalent dose from the alpha radiation is greater than the equivalent dose from the x-rays. In simple terms, there are various ways to arrive at a particular absorbed dose, and the biological response will vary according to both the absorbed dose and how the absorbed dose is achieved.

To calculate the equivalent dose from a particular radiation type, a multiplier called a radiation weighting factor (w_R) is introduced to convert absorbed dose to equivalent dose, H_T:

$$H_T = \sum w_R \cdot D_R \tag{2.15}$$

where D_R is the absorbed dose of the particular type of radiation. These weightings vary from time to time as new research is published. Table 2.2 shows radiation weighting factors as recommended in ICRP Publication 60,[1] which is often called ICRP60, along with current recommendations.[2,3] When a radiation field contains more than one type of radiation or energy, the absorbed dose should be divided into proportional blocks of given w_R to assess the equivalent dose from each type or energy of radiation, and then summed to give the total equivalent dose.

Prior to ICRP60 a different weighting factor, called the Q factor, was in use and related to the linear energy transfer (LET) of a radiation. This weighted absorbed dose was called the dose equivalent, H. In essence, the values of w_R are broadly compatible with the Q values.

Although equivalent dose has the same basic unit as absorbed dose, when referring to equivalent dose the unit is called the sievert (Sv). The purpose of this change of name is to emphasize the change in meaning from a purely physical quantity to a biological response.

TABLE 2.2

Radiation Weighting Factors

Radiation	Radiation Weighting Factor (w_R)—1990 Recommendations		Radiation Weighting Factor (w_R)—2007 Recommendations	
All photons	1		1	
All electrons and muons	1		1	
Neutrons	<10 keV	5	< MeV	$2.5 + 18.2e^{-[\ln(E)]^2/6}$
	10–100 keV	10	1–50 MeV	$5.0 + 17.0e^{-[\ln(2\,E)]^2/6}$
	100 keV – 2 MeV	20	> 50 Mev	$2.5 + 3.2e^{-[\ln(0.04\,E)]^2/6}$
	2–20 MeV	10		
	>20 MeV	5		
Protons (other than recoil protons) >2 MeV	5		2	
Alpha particles, heavy nuclei, fission fragments	20		20	

2.3.3.2 Effective Dose

An additional consideration when predicting the effect of a dose of radiation on a subject is the fact that there is a variation in the sensitivity of tissue types to radiation. For example, gonads are much more sensitive to radiation than skin. The result of this is the addition of a tissue weighting factor, w_T, which operates as a multiplier of the equivalent dose to obtain the effective dose, E:

$$E_T = \sum_T w_T \sum w_R \cdot D_R = \sum_T w_T \cdot H_T \qquad (2.16)$$

The convenience of tissue weighting factors is that they not only allow for the prediction of effects of partial body or whole organ irradiations, but the multiplier value can easily be updated with changing knowledge of radiobiology; however, one should always use the current values as recommended by the ICRP. Tissue weighting factors given in the 1990 and 2007 ICRP recommendations are shown in Table 2.3. A numerical calculation of uniform equivalent dose over the whole body should give the same answer as the effective dose, that is, the tissue weighting factors should be normalized to unity.

TABLE 2.3

Tissue Weighting Factors

1990 ICRP Recommendations		2007 ICRP Recommendations		
Tissue/Organ	w_T	Tissue/Organ	w_T	$\Sigma\, w_T$
Gonads	0.20	Bone marrow (red), colon, lung, stomach, breast, remaining tissues	0.12	0.72
Bone marrow (red)	0.12	Gonads	0.08	0.08
Colon	0.12	Bladder, esophagus, liver, thyroid	0.04	0.16
Lung	0.12	Bone surface, brain, salivary glands, skin	0.01	0.04
Stomach	0.12			
Bladder	0.05			
Breast	0.05			
Esophagus	0.05			
Liver	0.05			
Thyroid	0.05			
Skin	0.01			
Bone surface	0.01			
Remaining tissues and organs	0.05			

Like equivalent dose, the unit used for effective dose is the sievert, for the same reason of the expression of a biological response rather than a physical quantity.

Example 2.4

Q: An absorbed dose of 2.5 Gy from a beta source is received to the thyroid of a patient. Calculate the equivalent and effective doses to the thyroid using the weighting factors given in the ICRP recommendations of 1990 and 2007.

A: Tables 2.2 and 2.3 show that in both ICRP publications the radiation weighting factor for electron radiation is 1. Therefore, the equivalent dose under both publications is

$$2.5 \ Gy*1 \ (w_R)* = 2.5 \ Sv$$

The tissue weighting factor differs between publications. The effective dose under the 1990 recommendations is

$$2.5 \ Gy*1 \ (w_R)* \ 0.05 \ (w_T) = 0.13 \ Sv$$

and the effective dose under the 2007 recommendations is

$$2.5 \ Gy*1 \ (w_R)* \ 0.04 \ (w_T) = 0.10 \ Sv$$

2.3.4 Other Units and Quantities

2.3.4.1 Committed Dose, Committed Equivalent Dose, and Committed Effective Dose

Committed dose, committed equivalent dose, and committed effective dose arise from the intake of radioactive material into the body. In these cases there will be a time period during which the material will be undergoing radioactive decay within the subject, committing them to a radiation dose until the material either decays, is excreted from the body, or the subject dies. It is then said that the subject is committed to receive a radiation dose for the coming years—hence the term committed dose. For calculation purposes, if the time is unspecified it is implied to be 50 years for adults and 70 years for children. Committed dose must be included in any calculations of annual dose for a subject for radiation protection purposes. These units provide the means to combine exposures that are not uniform and involve differing radiation components and pathways into a single risk parameter for protection purposes of the individual.

2.3.4.2 Collective Equivalent Dose and Collective Effective Dose

Collective equivalent dose, S_T, and collective effective dose, S, relate to exposures of groups of people. The purpose of these quantities is to predict the total consequences of an exposure of a population, either from a single event or from a long-term situation in an environment. The collective dose is obtained as the sum of all individual doses over a specified period from a source. These quantities are instruments for improvement of procedures and are not used to evaluate epidemiologic risk. They assume a linear dose-effect relationship without a threshold, and so to avoid aggregation of very low individual doses over long time periods, limiting conditions regarding dose and exposure time are set.

References

1. ICRP, *1990 Recommendations of the International Commission on Radiological Protection*, ICRP Publication 60. Oxford UK: Pergamon Press, 1990.
2. Clarke, R., 21st century challenges in radiation protection and shielding: Draft 2005 recommendations of ICRP. *Radiation Protection Dosimetry*, 115, 10–15, 2005.
3. ICRP, ICRP Publication Number 103: Recommendations of the ICRP. 2007: Elsevier.

3

Radiation and Risk: Radiobiological Background

Tomas Kron

CONTENTS

Radiobiology is a large and active field of research. It is beyond the scope of the present chapter to cover it comprehensively. However, it is necessary to understand at least the basic principle of radiation damage to cells in order to appreciate the background for radiation protection. Radiation effects on living cells, and consequently radiation biology, go beyond the energy deposited in matter. This becomes obvious when considering that a total body dose of 10 Gy is a lethal dose to humans; however, the actual energy deposited in joules per kilogram body mass is very small and would only lead to a temperature rise of less than 0.01°C. Therefore, the biological effect of radiation is not thermal and is not caused by the amount of energy itself, but its deposition pattern and the molecules affected. The importance of going beyond the physical interaction of radiation with matter is also illustrated in Figure 3.1, which shows the timescale of radiation effects. The effects of radiation on a living organism may extend over its whole life.

The first section of the present chapter describes the effects of radiation on living cells. This is not an in-depth discussion of all effects but an attempt to present fundamental principles and illustrate orders of magnitude. This is followed by an introduction into the effects of radiation on humans, distinguishing between stochastic and deterministic effects. As cancer induction is the most important risk associated with ionizing radiation, the chapter features a section on carcinogenesis. The final section is concerned with risk estimates. This is where numbers will be presented and orders of magnitude discussed in order to appreciate the need for dose constraints and their value.

For radiation protection professionals appendix B of Report 60 of the 1990 recommendations of the ICRP is a very valuable resource.[1] The reader is also referred to the Web pages of the International Commission on Radiological Protection (ICRP; http://www.icrp.org/index.asp) and the United Nations Scientific Committee on the Effects of Atomic Radiation (UNSCEAR; http://www.unscear.org/unscear/en/index.html).

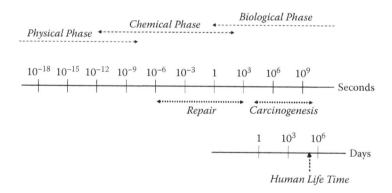

FIGURE 3.1
Time course of radiation interactions with biological material. Adapted from Steel.[2] The figure also indicates the approximate timescale for repair of DNA damage and the development of cancer.

3.1   Effects of Radiation on Cells

3.1.1   Timescale of Radiation Effects on Living Organisms

As shown in Figure 3.1, the radiation effects in biological matter can be classified into three different phases:

1. The physical phase, which includes the actual interaction of radiation with matter and the formation of radicals, which can cause indirect radiation damage to the cell.

2. The chemical phase, when lesions in the DNA may accumulate and enzyme reactions take place. Also, some fast and simple repair processes take place.

3. The biological phase, which encompasses the remainder of the repair processes, further cell divisions, mitotic death, apoptosis, and, finally, effect on organs and carcinogenesis.

3.1.2   What Happens in the First Pico Seconds after Irradiation of Cells?

The physical phase includes the actual interactions of radiation with matter, resulting in excitations and ionization events (compare with Chapter 2). Excitations typically consume the most energy, but it is ionization that leads to biological effects via the disruption of chemical bonds. The main target of importance for radiation effects in the cell is the deoxyribonucleic acid (DNA). Figure 3.2 shows the structure of the DNA schematically. It consists of a double helix with two sugar-phosphate strands that connect to each other through two types of base pairs: guanine (G) can bond with cytosine (C), and adenine (A) can bond with thymine (T). All four molecules are nucleic acids, with T and C being so-called pyrimidines (single-ring molecules) and G and A being purines (double-ring molecules).

DNA damage can occur in two different fashions: as direct interaction of radiation with the DNA or as indirect damage through radical formation, mostly in water. The relative contribution of the two mechanisms has been discussed at length in the literature—in experiments they are difficult to distinguish.[3] In practice the relative contribution of both will depend on the radiation quality and the environment of the DNA, in particular the level of oxygenation.

Excitations and ionization events occur on a timescale of femtoseconds (10^{-15} s) or less. Indirect radiation damage due to radical formation is several orders of magnitude slower (around 10^{-10} s). As biological systems consist largely of water, the predominant reaction is a breakdown of water molecules into a positively charged hydrogen ion (H^+; this is basically a proton) and an uncharged OH molecule. The OH molecule is highly reactive and therefore called a radical (often indicated by a dot next to the symbol $OH\cdot$). There are many different chemical reactions involving the products of radiation interaction with water. One of the more important ones is

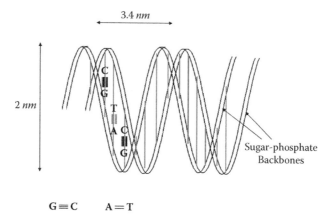

FIGURE 3.2
Schematic diagram of the DNA. G = guanine, C = cytosine, A = adenine, T = thymidine. Note that the bond between G and C is stronger than the one between A and T as it involves three hydrogen bonds as opposed to two in the case of A and T.

$$OH\cdot + OH\cdot \rightarrow H_2O_2 \tag{3.1}$$

Hydrogen peroxide (H_2O_2) is a reactive molecule with a relatively long lifetime. If these reactions are taking place close to the DNA, the radicals can diffuse to the DNA and react with it, thereby causing radiation damage, the indirect radiation damage mentioned above. The longer the lifetime of a radical, the further it can travel, which makes H_2O_2 an important radiation product for radiobiology.

Both direct and indirect radiation action on the DNA are much more complex than described here (compare Steel,[3] ICRP,[1] and UNSCEAR[4,5]). It is beyond the scope of the present chapter to describe this in depth; however, Figure 3.3 can illustrate some of the complexities arising just from the geometric formation of DNA. The human DNA molecule is some 2 m long. In order to pack this into the nucleus of a cell with a typical diameter of less than 10 μm, the molecule has to be packed rather tightly. As such, DNA has several superstructures that are illustrated in the figure. To appreciate both the importance of the DNA in a cell and the complexity of the damage formation within the DNA, it is useful to keep a number of facts in mind:

1. All molecules except for DNA are available in many copies in the cell. The DNA actually contains the "recipe" for the proteins in the cell—as such they can be resynthesized by expressing the appropriate gene.

2. The DNA is by far the largest molecule in the cell. It takes up approximately 10% of the space within the cell nucleus. The dimensions

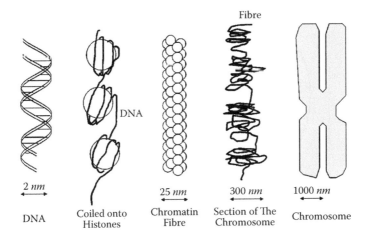

FIGURE 3.3
Structure of DNA: The DNA is coiled around histones that form nucleosomes. These are structured in chromatin fibers that form—in irregular spirals—the chromosomes. Adapted from Tubiana et al.[62]

are indicated in Figure 3.4. To make matters more complicated, DNA configuration will change throughout the cell cycle, where DNA is doubled in the S (synthesis) phase before dividing in the M (mitosis) phase. Figure 3.5 shows the cell cycle, which will be discussed in more detail below. Finally, even when not cycling, the DNA must change geometric conformation to allow transcription of genes.

3. Not all parts of the DNA are associated with genetic information. Figure 3.6 illustrates this.

Many of the processes of radical formation are enhanced in the presence of oxygen. This can explain the oxygen enhancement effect that is an important mechanism in the context of cell kill in radiotherapy. The oxygen enhancement ratio [OER = (cell kill in presence of oxygen)/(cell kill in hypoxic environment)] is of the order of three, which makes hypoxic cells much more difficult to kill.

Figure 3.7 shows some typical types of damage in the DNA. These include base damages, cross-links between different stands of DNA, protein cross-links, intercalation, and single and double strand breaks. The complexity of the possible processes can be seen by the fact that more than one hundred different types of base damage have been identified.[6] The number of single strand breaks and base damages per unit radiation is typically nearly two orders of magnitude larger than the number of double strand breaks. However, it was found that the number of double strand breaks correlates

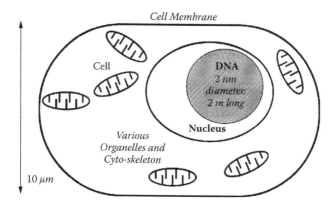

FIGURE 3.4
Approximation of dimensions in a mammalian cell; the DNA takes up a considerable amount of physical space.

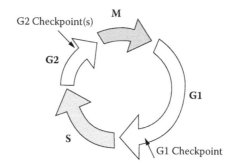

FIGURE 3.5
Simplified illustration of the cell cycle. G = gap or growth phases, S = DNA synthesis phase, M = mitosis or cell division phase. The G and S phases are often also collectively referred to as interphase.

best with cell kill[3]—as such, it is generally assumed that double strand breaks are the most significant lesion for cell kill.

Double strand breaks require a break in both strands of the DNA, about 2 nm apart. While lesions can travel a few base pairs along the DNA, this still requires two ionization events to be within a few nanometers of each other. The likelihood of this occurring depends on the type of radiation and the radiation dose. The research field concerned with these effects is called microdosimetry.[7-9]

Photons and high-energy electrons have a linear energy transfer (LET) of approximately 0.3 keV/μm, which is considered low. As the average energy lost by radiation for one ionization event is around 30 eV, one can find

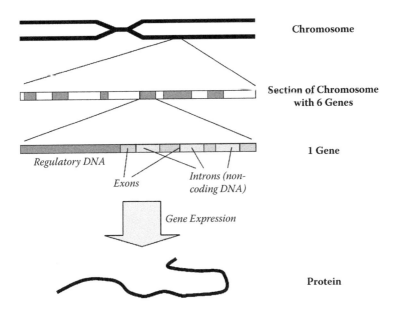

FIGURE 3.6
Chromosomes and genes. There are approximately thirty thousand genes encoded in human DNA. Each gene can be used to produce a specific molecule. If this happens, the gene is expressed.

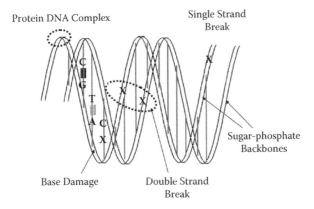

FIGURE 3.7
Different types of DNA damage due to radiation. Note that this illustration does not include all possible types of damage.

approximately ten ionizations per micrometer, which is about the dimension of a cell nucleus. Compared to the dimensions of the DNA, these are large distances, and the probability of causing two ionization events in one segment of the DNA is small. One can also appreciate that the probability

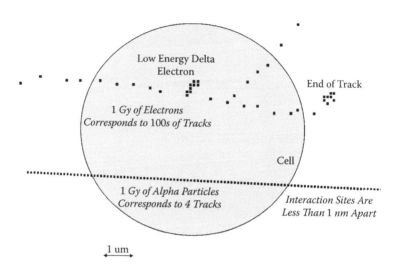

FIGURE 3.8
Ionization events in low- and high-LET radiation tracks illustrated on the scale of a single cell.

of a double strand break increases with radiation dose, as the likelihood of two radiation tracks coming close enough to each other increases. This is different for high-LET radiation such as alpha particles. Alpha particles (and other heavy charged particles) have a LET of approximately 200 keV/μm. Figure 3.8 illustrates this. In this case, two ionization events caused by the same particle are less than a nanometer apart, and the probability of a double strand break is consequently high.

From this simple geometric consideration one can deduct that radiation effects in cells will be dependent not only on the energy deposited (the radiation dose) but also on the LET. This is reflected in the radiation weighting factors discussed in Chapters 1 and 2.

3.1.3 Repair

Damage to the DNA is a common occurrence in nature. It may arise from ionizing radiation as well as other environmental agents such as chemicals (benzene, smoke), temperature, and UV light.

While the pattern of radiation damage is different for different agents (ionizing radiation produces a comparatively large number of double strand breaks), cells have developed mechanisms to survive despite DNA damage—independent of the type of damage. DNA damage repair is a complex process—there are at least 150 genes involved.[10] This demonstrates the importance of DNA repair mechanisms for living organisms. The following is a simplified discussion of the basic pathways. The reader is referred to two excellent recent reviews of DNA repair by R. Wood et al.[11] and E. Friedberg[12] for more details.

The repair process involves several basic steps:

1. Interaction of radiation with DNA yielding DNA damage (direct or indirect damage). Even under environmental conditions there will be ionization events. The annual dose from background radiation will result in nearly one hundred instances of DNA damage per cell per year. Only about 1% of them are double strand breaks.

2. DNA damage is sensed—there are several molecules involved in this process. One of the most important is the Ataxia telangiectasia gene product ATM (ATM stands for AT mutated).[13] This protein senses double strand breaks and initiates a response. As the name indicates, this gene is mutated in sufferers of Ataxia telangiectasia, a disease associated with a very high likelihood of cancer development.

3. Once damage is sensed, cells may go into cell cycle arrest to allow for more repair time. There are many checkpoints in the cell cycle with two important ones indicated in Figure 3.5. A checkpoint at the end of the G1 phase assesses if the cell is large enough and if there are sufficient nutrients to commence the synthesis phase of the cell cycle. For radiation damage, a checkpoint at the end of the G2 phase is particularly important. Prior to entering the cell division phase (M = mitosis) the cell checks if the cell is large enough to divide and if all DNA has been faithfully duplicated. If DNA damage is sensed, the cell will initiate a cell cycle arrest to respond to the damage.

4. Response to damage can have different forms—the most important ones are apoptosis and repair.

Apoptosis or programmed cell death is a mode of cell disintegration resulting in cell death.[14] It is characteristic for cells that sense they are damaged and are unable to repair. Apoptosis is a type of cell suicide that prevents DNA damage to be passed on to future cell generations. As such it is an important mechanism in tumor suppression. The tumor suppressor gene p53 is involved in regulating apoptosis.[15] It is therefore not surprising that many tumors have mutated p53 genes.

The type of repair mechanism depends on the type of damage. Single strand breaks, base damage, and many other simple lesions are repaired effectively by base excision repair. This is schematically illustrated in Figure 3.9. As one strand of the DNA is still intact, it can be used to replicate the second strand after the affected area has been excised by enzymes, which are proteins that facilitate specific chemical reactions.

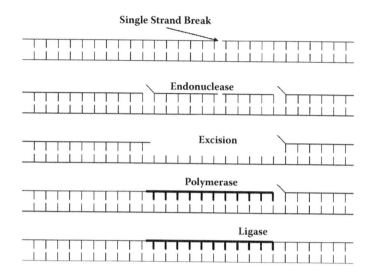

FIGURE 3.9
Base excision repair mechanism. Adapted from Tubiana et al.[62] It takes several separate enzyme reactions to excise the damage, resynthesize the broken strand of DNA using the second strand as template, and rejoin the DNA together.

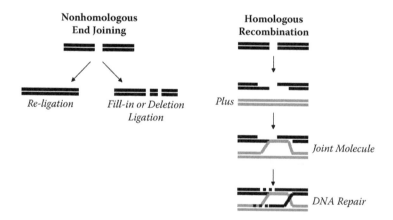

FIGURE 3.10
Mechanisms of double strand break repairs. Adapted from Friedberg.[12] Nonhomologous end joining is fast and rejoins the ends of the broken DNA. Loss of base pairs or insertion of others is not uncommon. Homologous recombination uses the second copy of DNA, which is available during the late S phase and G2 phase, to faithfully repair the damage using the intact copy.

Double strand breaks and other complex lesions are more difficult to repair. There are two basic mechanisms, which are illustrated in Figure 3.10:

- Homologous recombination (HR) makes use of the fact that in the S and G2 phases there are two copies of the DNA available. In this case, the undamaged copy can be used to facilitate a faithful repair. HR is relatively slow, only available in the S and G2 phases and virtually error free. It also varies more between different cell lines and is often defective in cancer cell lines. On the other hand, HR is more predominant in highly proliferate cells that go through the S phase more often. This makes sense in the context of early embryonic development where efficiency of repair is less important than correctness. HR is also responsible for the different radiation sensitivity of cells in different parts of the cell cycle. Cells are most sensitive to radiation in the M phase and most radiation resistant in the late S phase. However, it is important to note that there are differences in radiation sensitivity with phase of the cell cycle between different cell lines.[2]

- Double strand break joining or nonhomologous end joining (NHEJ) is independent of the cell cycle. As such it is predominant in nonproliferating cells. NHEJ uses re-ligation for "sticky" DNA ends after breakage. It may also result in deletions of some base pairs or the insertion of others. Overall, NHEJ is more error prone than HR but is in practice responsible for the majority (about 80%) of the repair of double strand breaks.

Both mechanisms involve several steps and protein complexes. It is important to note that the time required for repair will depend on the type of the lesion, the phase of the cell cycle, and the repair mechanism. As such, repair times can range from a few minutes to several hours. This will impact on the effect of time between irradiations (e.g., in fractionated radiotherapy) and the dose rate.

Misrepair of radiation damage is associated with radiation damage and cancer induction.[16] As such, it is important to study repair mechanisms and the probability of misrepair.

3.1.4 Bystanders, Genomic Instability, and Adaptive Responses

The discussion so far has neglected a number of observations that do not fit the rather simplistic model developed above. There are several effects on a cellular level, which modify the response of cells to radiation. Three important effects that may affect how we judge radiation effects for radiation protection purposes are:[17]

1. Bystander effect:[18,19] The term *bystander effect* describes the fact that chromosome aberrations and cell death have been observed in cells that have not directly been targeted by a toxic agent, such as radiation. A number of experiments have demonstrated that radiation does not need to deposit energy in the cell nucleus in order to produce cell death. This has been elegantly shown using charged particle microbeams.[18] They allow targeting individual cells or even parts of a cell. The "communication" of damage between cells has also been shown in radiation beams used for radiotherapy.[20] While the mechanism of the bystander effect is not entirely clear,[21,22] it may be important for the appreciation of the effects of low doses of radiation as the effective target cell population is increased.[17]

2. Genomic instability:[18,19] Genomic instability describes a delayed mechanism of radiation damage and challenges the notion that radiation damage is manifest in the cells that are actually irradiated. It has been observed that genetic mutations and chromosomal damage can occur in many cell generations after the actual irradiation. While this effect appears to be not present in all cell lines and its mechanism is not clear, it would again have implications for the interpretation of radiation damage in the context of radiation protection.

3. While the two previous effects tend to increase the damage observed after irradiation, adaptive responses are a mechanism that may prevent some of the damage.[23] The term describes the phenomenon that radiation effects are less pronounced if cells have been irradiated prior to the experiment using a small dose of radiation. Like the previous two effects, adaptive responses have not been observed in all cell lines,[17] but they may need to be considered when judging the potential damage due to irradiation.

3.2   Effects of Radiation on Humans

In order to determine risk estimates for radiation protection purposes the most relevant data are from radiation exposures of humans. Two different types of radiation effects can be distinguished: deterministic and stochastic. The difference between the two types is illustrated in Figure 3.11.

3.2.1   Deterministic Effects

Deterministic effects are generally due to cell killing—with the exception being cataract formation, discussed below.[24] They have a dose threshold below which no effect is observed. Above this threshold, the severity of harm increases with dose. The effect is specific for a particular type of tis-

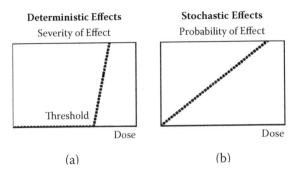

FIGURE 3.11
Deterministic and stochastic radiation effects. Note the difference in y-axis.

sue. Examples for deterministic effects are erythema of the skin and damage to nerve cells (e.g., leading to paralysis in the case of spinal cord damage). For most deterministic effects in humans, the threshold dose is of the order of several gray. The actual threshold dose also depends on dose delivery mode, with single exposures being typically more detrimental than protracted exposure. In general, deterministic effects are of primary concern at high radiation dose levels.

The effects of total body irradiation can be taken to illustrate this concept (compare Herrmann and Baumann[2] and Hall and Giaccia[25]):

- Radiation sickness may occur at doses around 2 Gy where a significant number of intestinal cells are killed.
- At higher doses the cells of the bone marrow that produce blood cells, and, therefore, support the immune system, are the most important consideration. Effects on the immune system can be observed at doses between 1 and 10 Gy. In total body irradiation for therapeutic purposes immunosuppression is actually the desired effect. However, these patients require intensive care, and for healthy individuals the exposure to several gray of radiation can lead to death due to the shutdown of the immune system within weeks of irradiation.[26] The delay is due to the relatively slow turnover of blood cells. Without intervention (antibiotics, transfusions, bone marrow transplant) this hematopoietic syndrome (3–8 Gy) leads to death in 3–8 weeks.
- Effects on the small intestine can be observed at doses between 5 and 20 Gy. Once the acute dose to the body exceeds 10 Gy, internal bleeding follows and death may occur within a few days due to GI tract depopulation of epithelium (gastrointestinal syndrome). The prognosis of these patients even with therapeutic intervention is not good.
- At even higher dose levels (20–50 Gy), the central nervous system is directly affected. The cerebrovascular syndrome (100 Gy) is characterized by an increase in small-vessel permeability and leads to

death in 1 to 2 days. There is no therapeutic approach that can significantly prolong life in these patients.

From a radiation protection point of view the hematopoetic syndrome is
most relevant as early detection and therapeutic intervention result in good
chances of survival. Radiation injury can be determined first from the lymphocyte count.[2] The count reduces within less than 1 day with a nadir after 3 to
6 days. The lymphocyte count reduction is proportional to the severity of the
radiation injury and can as such be used as a biological radiation dosimeter.

Another important (and often discussed) deterministic effect is cataract
formation. A cataract is defined as an opacification of the normally transparent lens of the eye. As a radiation effect it constitutes a failure of the pre-
equatorial epithelial dividing cells to differentiate properly, an effect that
has been reported to occur after single exposures of 2 Gy or a fractionated
exposure of 10 Gy. With increasing dose, the latency period decreases.[25]

3.2.2   Stochastic Effects

Stochastic effects are considered to be applicable to all dose levels. The most
important stochastic effect from a radiation protection point of view is cancer induction, which will be discussed in more detail in the next section. At
least from a conceptual point of view, a single radiation-induced event can
lead to cell changes in the DNA, which can result in transformation of a normal cell to a malignant cell. In this case, no threshold dose exists, as shown
in Figure 3.11b. The severity of the effect is independent of the dose as cancer
induction is equally detrimental if induced by small or large radiation doses.
However, the probability of the effect increases with dose.

Cancer induction is not the only stochastic effect. Much of the risk analysis for stochastic effects is derived from the survivors of the nuclear bomb
explosions in Japan in 1945.[27] The data have also been analyzed to determine the noncancer mortality for those who died between 1950 and 1990.[28]
A statistically significant increase with radiation dose has been shown for
stroke, heart, respiratory, and digestive tract diseases at equivalent dose levels above about 1 Sv.

The difference between deterministic and stochastic radiation effects can
be compared with the hazards of driving a car. The number of hours behind
the wheel is a stochastic parameter—the more one drives, the more likely it
is that one is involved in an accident. The severity of the accident, however,
is independent of the number of hours driven. Getting a ticket for speed-
ing, on the other hand, may be considered a deterministic effect—there is a
threshold (the speed limit) and the severity of the penalty increases with the
excess speed.

In the context of stochastic and deterministic effect, radiation protection
has consequently two different objectives:

1. Prevention of deterministic effects (except in radiotherapy, those that are intentionally produced, but including those that are not intended, such as accidental medical exposure)
2. Reduction of the probability of stochastic effects and ensuring that risks associated with stochastic radiation events are acceptable

3.2.3 Hereditary Effects and Irradiation in Utero

As radiation interacts with the DNA it is important to also consider effects where damage is passed on to future generations.[5] These effects are often considered separately from the stochastic effects that express themselves in the exposed individual. While intensive studies of children of the atomic bomb survivors have failed to identify an increase in congenital anomalies, cancer, chromosome aberrations in circulating lymphocytes, or mutational blood protein changes,[29] there is ample evidence of ionizing radiation causing heritable mutations in many plants and animals. Therefore, it must be assumed that heritable mutations occur also in humans.

As cells divide rapidly in an embryo during gestation, one can expect that radiation damage will express itself early and embryos are more radiation sensitive than adults. In addition to this, they have a larger life expectancy throughout which adverse effects may occur. The International Commission on Radiological Protection (ICRP) has two recent reports that address these issues: *Radiological Protection and Safety and Pregnancy*, ICRP Report 77,[30] and *Biological Effects after Prenatal Irradiation (Embryo and Fetus)*, ICRP Publication 90.[31]

Not surprisingly, the radiation effects on embryos in prenatal exposure depend on the time of gestation, and three main phases can be distinguished:[2]

1. In the preimplantation period (0 to 9 days) exposure typically leads to death of the embryo: Due to the small number of cells, the fetus is very radiation sensitive throughout this phase and a dose of 1 Gy is considered to kill half of all embryos. Growth retardation and malformations are typically not seen in this phase of development.
2. In the organogenesis/embryonic period (10 days–6 weeks) malformation of organs, small head size, and intrauterine growth retardation are observed.
3. In the early fetal period (6–25 weeks) mental and growth retardation is the most significant effect. Also, the nervous system may be affected. Radiation sensitivity is decreasing throughout the fetal period; however, increased carcinogenesis must be expected as a result of radiation exposure throughout the fetal phase. Carcinogenesis due to diagnostic exposures of women late in pregnancy has been shown to be likely without a dose threshold as would be expected from stochastic effects.[32]

Radiation effects in humans are also categorized as *early*, within 1 year of irradiation, and *late*, appearing after this period. Somatic effects are those that affect the human body, including embryo, whereas genetic effects are due to chromosome mutations that produce heritable effects. If a recessive gene is affected, then mutation of offspring will only be apparent if two recessive genes are linked and the recessive trait shows through.

It is important to realize, therefore, that an adverse radiation effect is associated not only with a certain dose but also with a certain latency period. This is often indicated in a notation such as $LD_{50/30}$, which means that 50% of a study population die (LD = lethal dose) within 30 days.

3.3   Cancer Induction

As cancer induction is the most significant risk for humans associated with the relatively low doses and dose rates encountered in radiation protection, it is important to discuss the characteristics of cancer prior to looking into the dose-response relationship of cancer induction.

3.3.1   What Defines Cancer?

Cancer is a genetic disease. It is characterized by six hallmarks that have been described by Hanahan and Weinberg.[33] Figure 3.12 summarizes them.

- Evading apoptosis is a feature of most cancers and resistance to it has been linked to tumorogenesis.[34] Apoptosis is a process that helps cells to balance cell proliferation and cell death.[35] As tumor growth depends on both the growth rate and the cell loss, the avoidance of cell death due to apoptosis (a very important mechanism for cell loss) will enhance the tumor growth. Apoptosis has a typical appearance: the cell membranes rupture, the cellular skeleton breaks down, and the nucleus with its DNA is fragmented. All this happens after apoptosis is triggered, within less than 2 hours, and the remains of the cell are easy to "be swallowed" by macrophages. The most common way that cells can acquire resistance to apoptosis is via mutation of the tumor suppressor gene p53.[36] This appears to be present in more than 50% of human cancers.[33]
- Self-sufficiency in growth stimulation is another characteristic of tumor cells. Normal cells require a stimulus, such as a growth factor, to grow. The stimulus is typically received by a growth factor receptor protein in the cell membrane. As such, growth factor

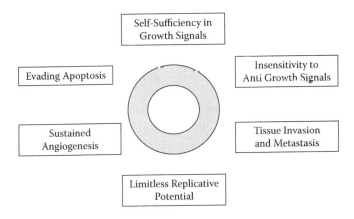

FIGURE 3.12
The hallmarks of cancer. Adapted from Hanahan and Weinberg.[33]

receptors are good targets for cancer drugs, with cetuximab being one important example.[37] Tumor cells often express more growth factor receptors and may also excrete growth factors for self-stimulation of growth.

- Insensitivity to growth inhibition is the correlate to self-sufficiency in growth stimulation. In this case the antigrowth signals that maintain cell homeostasis are blocked.
- Limitless proliferation (immortality). The previous three hallmarks of cancer cells de-couple their growth from the cell environment.[33] However, in normal cells infinite proliferation is usually not possible. At each cell division the ends of DNA in linear chromosomes (telomers) get a bit shorter.[38] This process limits the total number of cell divisions (to something of the order of sixty to seventy[33]) and produces senescence in cells. Tumor cells have acquired a mechanism to maintain telomer length and as such can be considered immortal.
- Sustained angiogenesis is required to provide nutrients and oxygen to a growing tumor. Once a tumor reaches a size of 0.1 mm diameter it cannot be adequately provided for by external blood vessels. As such, tumor cells express angiogenic factors (such as vascular endothelial growth factor [VEGF]) that lead to the development of new blood vessels.[39] This is an essential process for the growth of a tumor to macroscopic size and eventually to metastasis. Consequently, angiogenesis is a promising target for cancer therapy, and the first new drugs have commenced clinicaltrials.[40,41]

- Tissue invasion and metastasis is the final hallmark of cancer. About 90% of cancer deaths are actually due to metastases.[42]

The awareness of these hallmarks of cancer is important for radiation protection considerations from a number of perspectives:

- All these characteristics are associated with changes in the DNA, and it is evident that several events are required to cause the mutation of a normal cell to a cancer. In practice, this transformation is therefore not a single event process and consists of many steps. Hanahan and Weinberg[33] discuss several pathways how a normal cell can acquire all the hallmarks of cancer. They identify at least five separate steps. This is mirrored in the more traditional model of three necessary steps that transform a normal cell into a cancer cell: initiation, promotion, and progression.[43] In this model radiation is thought to be particularly important for the progression stage.
- In mathematical models of cancer induction for risk analysis it will make a large difference as to which step in the process radiation is responsible for.[16]
- The identification of genetic trades of cancer induction may allow learning from other diseases that are accompanied by increased susceptibility to cancer induction. Ataxia telangiectasia with its defect repair mechanism was mentioned already in Section 3.1.3. Other diseases and their potential link to radiation damage are summarized by ICRP Report 79.[44]
- As discussed in the next section, epidemiological data are very difficult to interpret. It is of course also unethical to perform radiation experiments with humans. As such, it is essential to study the molecular pathways of radiation effects in order to determine the dose-risk relationship more accurately.[6,16]

Many of these biological data are new, and there is a very rapidly developing body of knowledge about the biology of cancer development. However, it will still take many more years before these data can be used to calculate risks from exposure of humans to ionizing radiation. Therefore, most radiation protection considerations are still based on data obtained from humans.[45]

3.3.2   Epidemiology

Good epidemiological data on radiation exposure of humans are hard to find. The problems are obvious:

1. Typical radiation doses received in the environment or from common diagnostic procedures are small. Consequently, the risk of any effect is also small, and very large numbers of persons are required to observe even a few effects. The size of a cohort required to detect the additional cancer mortality from an exposure to 50 Sv is of the order of a hundred thousand exposed persons.[46]

2. Latency periods (and therefore observation periods) are quite long (many decades).

3. Radiation exposure is often protracted over a long period and accompanied by other confounding factors such as socioeconomic status and smoking habits.

4. Radiation sensitivity may vary among persons, and a small group with high radiation sensitivity can dominate the observations at low radiation doses.[46]

5. For ethical reasons no prospective studies can be performed. As such most data are retrospective.

6. Data often involve accidents and classified information. The reactor accident at Chernobyl, where there is still a debate about the true number of deaths associated with it, is an example of this.[47]

7. Dosimetry is typically very difficult and may involve mixed fields such as gamma and neutron radiation in the case of the survivors of the nuclear bombing in Japan.

8. It is often difficult to translate effects from populations studied at different times in different locations to other groups. For example, no one would try to explain cardiovascular disease in modern North America from data gained 60 years ago in Asia.

However, due to the importance of the question of radiation effects on humans a large amount of data have been collected over the years—in fact, ionizing radiation is one the best-studied carcinogens.[45] Several important data sets are available as explained in an excellent review by E. Cardis et al.:[45]

- The most important study group is the survivors of the nuclear bombing of Hiroshima and Nagasaki at the end of World War II in 1945. This group of some hundred thousand persons has been followed up closely over more than 50 years, and it forms the main basis for radiation risk estimates.[4] Also, the recommendations of the ICRP are largely based on this study group.[1] The advantages of this study are that a relatively large cohort was followed closely over more than 50 years. However, the acute nature of the exposure does not match most radiation protection scenarios, and there has been ongoing discussion about the dose estimates to individuals.[1,45]

- Radium dial painters are another well-studied group.[48] This experience concerns 820 women employed in the watchmaking industry before 1930 to paint numbers on watch faces. The luminescent paint used for this contained significant amounts of radium that was partially ingested when the tip of the paintbrush was sharpened with the workers' lips. It was found that the radium, which is deposited in bone, caused bone sarcomas and head and neck tumors.

- There is a large database of workers in the nuclear industry. Again, the size of this group is of the order of a hundred thousand persons with well-documented radiation exposure records.[45] There are some limitations to the study population, as nuclear workers typically are well paid and health checked. There is a large body of literature about these exposures, which is reviewed by Cardis et al.[45] The most important finding was a dose-related increase in leukemia in the workers. It can be expected that more workers from many countries will be included in the study population, leaving the group a very relevant cohort for the study of the effects of low protracted doses of radiation.

- Victims of the nuclear accident at Chernobyl in 1986[49] are a more recent large group of persons exposed to ionizing radiation. Of particular relevance are the so-called liquidators, consisting of persons involved in the cleanup after the accident. This group comprises hundreds of thousands of persons with a wide range of different exposures, from a few milligray to several gray. The other group of interest is children exposed to 131-iodine via inhalation after the accident. Overall, more than 800 cases of thyroid cancer due to exposure to ^{131}I have been reported in Belarus alone.[45] In this context it is important to note that many statistics only report fatalities; in the case of thyroid cancer, this will show quite a different picture than the actual cancer incidence, as 95% of thyroid cancers can be cured. This observation highlights the fact that detriment from radiation exposure may not always be death.

- Patients undergoing medical exposures for diagnosis and therapy are increasingly being studied for development of detrimental effects due to radiation. In particular, secondary cancers due to radiotherapy have become a topic of discussion in the context of an increase in low dose to large areas of the patient's body due to modern radiotherapy delivery techniques.[50] As many of the radiotherapy patients are now long-term survivors, secondary cancers due to radiation exposure during treatment have been observed. Of particular interest are patients undergoing total body irradiation for bone marrow transplants, lymphoma treatments, and adjuvant breast cancer treatment. In many of these treatments patients are relatively young or are treated in an adjuvant setting where risks and benefits must be evaluated differently.

In addition to this there are several studies comparing groups of persons living under different environmental conditions. Of particular interest are observations on environmental irradiation and cancer in India[51] and an extensive study of the correlation between lung cancer and the concentration of radon in dwellings.[52] Both studies did not demonstrate an adverse effect of exposure to radiation. They both actually demonstrated that radiation was associated with a benefit for persons receiving higher radiation doses. While none of these studies can demonstrate a causal connection between dose and reduced cancer incidence, the data illustrate the complexity of the issue and the difficulty of evaluating epidemiological studies of large numbers of persons exposed to small radiation doses. It is beyond the scope of the present chapter to discuss this and other evidence of beneficial effects. However, it appears that it is not possible to consider all cancers the same, and for some (e.g., lung cancer and radon exposure) there may even be a positive effect of radiation.

Development of breast cancer can be considered an example for some of the issues faced. As the breast is a particularly sensitive organ for the development of secondary cancers after radiotherapy, numerous studies have looked into the risk of secondary breast cancer after radiotherapy and other exposures. A number of these studies are summarized in Table 3.1, together with other data and dose-response estimates from the literature. A summary of the literature is also provided by UNSCEAR[4] in its Annex I.

The data of Table 3.1 are also plotted in Figure 3.13 for illustration purposes. The figure helps to illustrate the difficulty of making exact predictions about radiation risk. While there is a general trend to an increased breast cancer risk at high radiation doses, the scatter of the data makes the interpretation difficult. This variability is due to a variety of sources, such as:

- Uncertainty in dose delivered (e.g., in radiotherapy studies this involves scatter and transmission)
- Variation of the doses actually received by individuals in the study population
- Variations in exposure pattern with time
- Follow-up time
- Competing risks
- Differences in study population (ethnicity, general health, and, most importantly, age at exposure)

3.4 Risk Estimates

For radiation protection purposes it is essential to quantify risks associated with ionizing radiation. There is no doubt that at high radiation doses there

TABLE 3.1

Risk of Breast Cancer as a Function of Estimated Equivalent Dose Received

Dose (mSv)	Relative Risk	Reference	Comments
2	1		Natural background
10	1.14	Land et al.[64]	Exposure prior to age 35; data from A-bomb survivors
30	1.87	Pukkala et al.[65]	Study on Finnish airline cabin attendants; dose estimates affected by uncertainty
100	1.36	Miller et al.[66]	Estimate from fluoroscopy data of more than 30,000 tuberculosis patients; estimate compares exposures greater than 100 mSv with those less than 100 mSv; highest risk when exposed early in life
100	1.8	ICRP[1]	From Table B.9 based on acute whole body uniform irradiation with low-LET radiation; the risk is considerably higher if irradiation occurs early in life; life expectancy as per U.S. population
130	1.8	Hoffman et al.[67]	Study on patients irradiated on multiple occasions in adolescence due to scoliosis
290	1.2	Lundell et al.[68]	Swedish study on 17,000 women irradiated as children for skin hemangioma
500	1.18	Clarke et al.[69]	Risk of additional cancers in contralateral breast in patients who had received radiotherapy for breast cancer—Early Breast Cancer Trials' Collaborators Group; meta-analysis of 42,000 women in 78 clinical trials, mainly older women
1,000	3.41	Carmichael et al.[70]	Summary of data from nuclear bomb survivors and patients receiving radiotherapy; irradiation prior to age 10
1,000	2.25	Carmichael et al.[70]	As above; irradiation at age 10–30
1,000	1.48	Carmichael et al.[70]	As above; irradiation at age 30–50
4,000	3.2	Travis et al.[71]	Estimates from patients treated for Hodgkin disease with radiotherapy and chemotherapy
40,000	8	Travis et al.[71]	Estimates from patients treated for Hodgkin disease with radiotherapy and chemotherapy
40,000	18.1	Bhatia et al. (1996)[72]	Estimates after irradiation for Hodgkin disease

Note: The table was compiled from various sources and the equivalent dose and relative risk calculated from the published data. There is considerable room for interpretation, and uncertainties exist in particular as the study populations vary. As such, the data in the table must be seen as an illustration only.

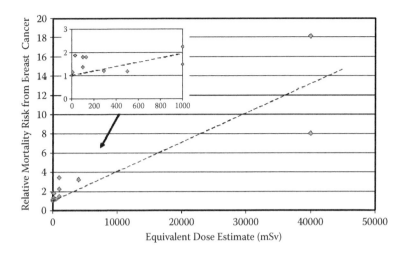

FIGURE 3.13
Relative risk of developing breast cancer after irradiation. The data plotted are taken from various sources in the literature, shown in Table 3.1. The line shows the best linear fit to the data with the fit forced through 1 at no dose ($r^2 = 0.784$). The highest doses were given in many fractions as part of radio therapy.

is a risk of cancer induction by radiation. However, it is important to quantify the risks for particular radiation types and organ exposures in order to evaluate alternative techniques in terms of their risk. Also, for regulatory purposes it is essential to determine the risks at the low-dose levels to which we may be exposed. A review of risk estimates for radiation protection is given by the National Council on Radiation Protection and Measurements (NCRP) in the United States,[53] and the ICRP has devoted its recent Report 99 particularly to the risks due to low levels of radiation.[35]

3.4.1 Orders of Magnitude

A reasonable way to commence quantification of risks is to establish quantitative measures for radiation effects on living cells. This is attempted in Figures 3.14 to 3.16.

Figure 3.14 shows the orders of magnitude of radiation effects on three different systems, a single cell, a group of 1 million cells, and a human person with about 10^{14} cells. It becomes evident that it is not easy to translate individual events on a molecular or cellular scale into biological effects that are meaningful from a radiation protection perspective. However, the numbers shown are self-consistent, as the example calculation at the end of the chapter shows.

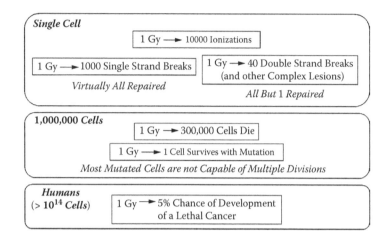

FIGURE 3.14
Orders of magnitude for radiation effects on three different levels for 1 Gy absorbed dose.

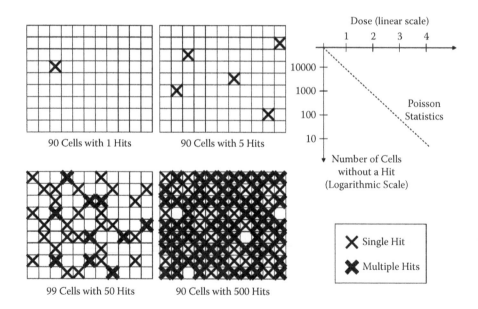

FIGURE 3.15
Graphical illustration of the Poisson statistics for cell kill due to a single hit: a single hit in an ensemble of cells will always kill one cell. The probability of hitting a cell is purely random; therefore, the likelihood increases that a cell is hit twice when the number of hits increases. This "overkill" will not result in larger cell kill, even if the number of radiation events exceeds the number of cells in an ensemble.

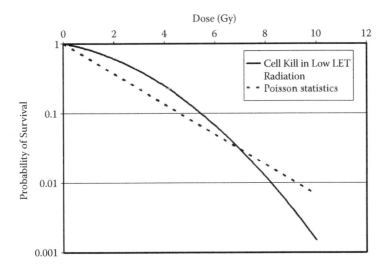

FIGURE 3.16
Cell kill as a function of dose. Comparison of the single-hit Poisson statistics with cell kill observed in cell survival assays.

3.4.2 Models of Cell Kill

A different problem with these interpretations is illustrated in Figures 3.15 and 3.16. Figure 3.15 shows a mechanistic model for hits in a cell. If an effect is due to a single hit, then only a cell where no radiation has hit the DNA can survive. In an ensemble with many cells the first hit will always incapacitate one cell. The more hits there are in the cell ensemble, the more likely it is that the same cell is hit more than once. In this simple model this would constitute a "waste" of radiation, as a cell can only die once. The statistics that determines the number of surviving cells, N, in an ensemble of n cells hit by m ionization event in DNA is called Poisson statistics:

$$N (n, m) = n\, e^{-m} \tag{3.2}$$

In this case N is mathematically the number of cells with no event. It is interesting to note that even if a very high radiation dose is delivered to a group of cells (such as a tumor in radiotherapy), there is always a small but finite chance that one cell will not be hit. This type of dose-response describes high-LET radiation reasonably well. However, at low LET the dose-response curve for cell survival after irradiation shows a distinct shoulder, as shown in Figure 3.16. There are several ways to explain the shoulder, which are discussed in radiobiology books such as Steel,[3] Herrmann and Baumann,[2] Wigg,[54] and Hall and Giaccia.[25] These include:

- The requirement of more than one hit for cell kill (the Poisson statistics in this case will yield a shoulder curve)
- The existence of more than one target in the cell
- The fact that repair is less effective at higher radiation doses, which will make the survival curve bend down

It is beyond the scope of the present chapter to discuss these theories and the supporting evidence in detail. In general, the shoulder curve in Figure 3.16 can be described by a linear quadratic formula:

$$S = n \exp(-(\alpha D + \beta D^2)) \qquad (3.3)$$

with S as the cell survival, n the number of cells, and D the radiation dose received. The parameters α and β are chosen to describe the curve. The linear quadratic model of equation (3.3) is widely used in radiotherapy and can be extended to include fractionation,[55] protracted irradiation, and overall treatment time.[56] The parameter α determines the initial slope of the curve at low doses, while the ratio between α and β describes the shoulder of the curve. The parameters α and β are characteristic for a particular tissue (or tumor) type, and the α/β ratio is one of the most important parameters used for radiobiological modeling in radiotherapy.[54] Due to its usefulness there is a wealth of literature about the linear quadratic (or "alphabeta") model. The reader is referred to radiobiology texts such as Steel,[3] Herrmann and Baumann,[2] Wigg,[54] and Hall[25] for more details.

3.4.3   Linear No-Threshold Model

One of the most important problems in radiation protection is to quantify the effects of the small doses of radiation exposure typically received in the context of uses of ionizing radiation in modern society.[35] Detrimental effects of radiation due to the stochastic effects applicable to low-level radiation are indistinguishable from detrimental effects due to other sources. It is at present impossible to identify exactly what has caused a particular cancer—it may be genetic, from environmental or hormonal influences, or from radiation. As discussed in Section 3.1, it is actually likely that several of these factors must come together to cause a cancer. Therefore, it is difficult to determine the small number of additional events due to radiation exposure on the background of the natural incidence of cancer. This is illustrated in Figure 3.17, which shows the death rate from cancer as a function of radiation dose.

The right-hand side of the curve at dose levels above about 100 mSv is well established and documented. It is mostly based on the experience gained of the survivors of the nuclear bombs in Japan. These data are augmented by other epidemiological information obtained in medical exposures and from workers in nuclear power plants. The data have been analyzed by the ICRP and a risk of about 10% per sievert has been established for cancer mortality

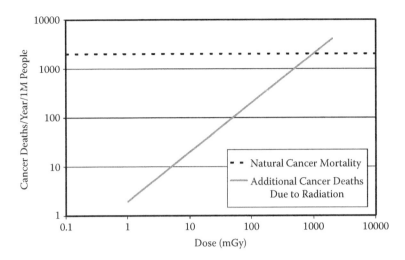

FIGURE 3.17
Cancer mortality in humans. Shown is the death rate from natural cancer incidence compared to the additional number of cancer fatalities observed as a function of total body radiation dose. Note that both axes of the plot are logarithmic.

for the whole population for high doses and low-LET radiation.[1] This overall risk can be broken down to contributions of specific organs. The largest contributions are from stomach, intestines, and lung followed by bone marrow. This experience is reflected in the tissue weighting factors recommended by the ICRP and reproduced in Table 2.3.

In order to determine the effect at lower doses it is necessary to extrapolate the dose-effect curve from high exposures to low ones, as indicated in the figure. It is difficult to determine the mathematical relation that should be used for the extrapolation. The simplest method is a linear extrapolation; however, the ICRP also notes that many effects attributed to ionizing radiation exhibit a curvilinear behavior.[1] As such, the commission has introduced a dose and dose rate effectiveness factor (DDREF), which takes into consideration that low doses delivered at low dose rates over long periods of time are less effective in producing cancer than the high acute doses for which observations are available. The DDREF is to be applied when the dose received is less than 0.2 Gy, or if a higher dose is received, if it was delivered with a dose rate less than 0.1 Gy per hour.

The ICRP in Report 60 recommends the use of a DDREF of 2. It is important to note that this factor is not applicable to high-LET radiation. Therefore, the risk of contracting a fatal cancer from exposure to low-LET radiation delivered at low dose rates is 5% per Sv in the whole population. For radiation workers the average age is higher; therefore, a figure of 4% per Sv should be used.[1]

A few things need to be kept in mind when using the linear extrapolation in the linear no-threshold (LNT) model:

- The linear extrapolation implies (even with a DDREF) that every dose received is potentially harmful. This is reflected in the radiation protection principle of justification. No dose shall be delivered if this is not balanced by appropriate benefits.
- The model is only applicable to stochastic effects.
- Different effects and different organs may have a different dose-response relationship; however, the LNT model is considered a reasonable summation of all effects.

3.4.4 Estimating Risk

The linear no-threshold model described above forms the basis of our current system of radiation protection.[1,57–59] The linear no-threshold model is not uncontroversial (compare point/counterpoint discussion in Strom et al.[60]), and it has been debated widely in the literature (compare, for example, Charles[61] and Tubiana et al.[62] and references therein). It is generally acknowledged that it represents a simplification—albeit a useful one. Other aspects that could be taken into consideration are:

- The magnitude of hereditary effects that will affect future generations. It is now generally assumed that 1 Gy of radiation will cause genetic mutations in just under 0.5% of the population.[5,17] This constitutes only a small fraction of the natural occurrence of the same disorders in the human population.
- Hormesis is the term used to describe beneficial effects from exposure to low doses of ionization. Several of these have been observed, such as the impact of radon on lung cancer[52] or the general health benefits observed in Taiwan after ^{60}Co accidentally contaminated recycled steel used in the building industry.[63] Some of these effects have been attributed to a stimulation of the immune or cell repair system at low doses. However, hormesis is observed in some cell lines and not in others, and it is often difficult to define what constitutes a benefit.

In conclusion, it is important to consider that a system of risk assessment for radiation protection purposes must be simple and robust. While it is important to consider all different effects and the complexity of radiation effects in particular at low doses, it is fair to summarize the experience as the ICRP does it in its recent Report 99:[35]

> The LNT (linear no-threshold) hypothesis, combined with an uncertain DDREF (dose/dose rate effectiveness factor) for extrapolation from high

doses, remains a prudent basis for radiation protection at low doses and low dose rates.

Example Calculation

A new tool in radiotherapy delivery is the possibility to acquire daily CT scans of patients prior to delivery of their treatment. This helps to position the treatment beams on a daily basis. The dose received by the volume imaged by the CT scan is 30 mGy (compare with Chapter 6), and overall thirty-three fractions of radiation are given to the thorax of a patient. The aim is to determine (a) the number of ionizations and double strand breaks per cell and (b) the overall additional risk to the patient.

a. The volume imaged an overall dose of 1 Gy is delivered in low-LET radiation. This leads to an energy deposition of 1 J, or 1.6×10^{19} eV, per kg of body mass. About 30 eV is on average expended for one ionization event, and as such there are about 5×10^{17} ionizations per kg, or 500,000 ionizations per cell (about 10^9 cells per g). In low-LET radiation the energy is deposited in approximately one thousand tracks per cell, which will feature five hundred ionizations each. Except for the very end of the tracks, therefore, the average spacing of ionizations is 0.02 μm assuming a cell diameter of 10 μm (LET = 0.15 keV/μm). The DNA is coiled in the cell nucleus—the volume occupied is of the order of 2 μm diameter. Therefore, we have 100 ionizations in the DNA per track, or overall a few thousand ionizations arising from all tracks traversing the DNA. This is compatible with the one thousand single strand breaks and thirty double strand breaks mentioned in Section 3.1.

b. The risk can be calculated from the equivalent dose and the organs involved. As the irradiation involves low-LET radiation (kilovolt x-rays), the radiation weighting factor is 1 and the equivalent dose has the same numerical value as the absorbed dose. In the case of the thorax the body parts irradiated are the breast, the lung, and some parts of the remainder (including the heart). From Table 2.3 in Chapter 2 we can see that the tissue weighting factors are 0.05, 0.12, and 0.05 for breast, lung, and remainder, respectively. Including a DDREF of 2 for low-dose-rate irradiation (<0.1 Gy/h) and assuming a risk of 5% per Sv for whole body irradiation, we can conclude that the additional risk of dying from a secondary radiation-induced cancer due to the diagnostic exposures during radiotherapy treatment is of the order of 1%. It is important to note, though, that this risk applies to the whole population and may be considerably lower in the case of an elderly patient.

Acknowledgment

The author acknowledges the ESTRO course on molecular oncology for some of the thoughts in this chapter and the motivation to learn more about radiobiology.

References

1. International Commission on Radiological Protection, *1990 Recommendations of the International Commission on Radiological Protection*, ICRP Report 60, Oxford: Pergamon Press, 1991.
2. Steel, G., *Basic Clinical Radiobiology*, 2nd ed., London: Edward Arnold, 1997.
3. Herrmann, T., and Baumann, M., *Klinische Strahlenbiologie*, 3rd ed., Jena, Germany: Gustav Fischer, 1997.
4. United Nations Scientific Committee on the Effects of Atomic Radiation, *Sources and Effects of Ionizing Radiation: 2000 Report*, New York: United Nations, 2000.
5. United Nations Scientific Committee on the Effects of Atomic Radiation, *Hereditary Effects of Radiation*, Report to the General Assembly with Scientific Annex, New York: United Nations, 2001.
6. Preston, R. J., Radiation biology: Concepts for radiation protection, *Health Phys.*, 87, 3–14, 2004.
7. Rossi, H., Microscopic energy distribution in matter. In *Radiation dosimetry*, vol. I, *Fundamentals*, edited by H. Attix, W. Roesch, and E. Tochilin, 43–92. London: Academic Press, 1968.
8. Rossi, H., and Zaider, M., Elements of microdosimetry, *Med. Physics*, 18, 1085–92, 1991.
9. International Commission on Radiation Units and Measurements, *Microdosimetry*, ICRU Report 36, ICRU, Bethesda, 1983.
10. Wood, R. D., Mitchell, M., and Lindahl, T., Human DNA repair genes, *Mutat. Res.*, 577, 275–283, 2005.
11. Wood, R. D., Mitchell, M., Sgouros, J., and Lindahl, T., Human DNA repair genes, *Science*, 291, 1284–89, 2001.
12. Friedberg, E. C., DNA damage and repair, *Nature*, 421, 436–40, 2003.
13. Bakkenist, C. J., and Kastan, M. B., DNA damage activates ATM through intermolecular autophosphorylation and dimer dissociation, *Nature*, 421, 499–506, 2003.
14. Kaina, B., DNA damage-triggered apoptosis: Critical role of DNA repair, double-strand breaks, cell proliferation and signaling, *Biochem. Pharmacol.*, 66, 1547–54, 2003.
15. Lowe, S. W., Schmitt, E. M., Smith, S. W., Osborne, B. A., and Jacks, T., p53 is required for radiation-induced apoptosis in mouse thymocytes, *Nature*, 362, 847–49, 1993.
16. Cox, R., Resolving the molecular mechanisms of radiation tumorigenesis: Past problems and future prospects, *Health Phys.*, 80, 344–48, 2001.
17. Preston, R. J., Radiation biology: Concepts for radiation protection, *Health Phys.*, 88, 545–56, 2005.

18. Morgan, W. F., Non-targeted and delayed effects of exposure to ionizing radiation. I. Radiation-induced genomic instability and bystander effects in vitro, *Radiat. Res.*, 159, 567–80, 2003.
19. Morgan, W. F., Non-targeted and delayed effects of exposure to ionizing radiation. II. Radiation-induced genomic instability and bystander effects in vivo, clastogenic factors and transgenerational effects, *Radiat. Res.*, 159, 581–96, 2003.
20. Suchowerska, N., Ebert, M. A., Zhang, M., and Jackson, M., In vitro response of tumour cells to non-uniform irradiation, *Phys. Med. Biol.*, 50, 3041–51, 2005.
21. Morgan, W. F., Is there a common mechanism underlying genomic instability, bystander effects and other nontargeted effects of exposure to ionizing radiation? *Oncogene*, 22, 7094–99, 2003.
22. Chaudhry, M. A., Bystander effect: Biological endpoints and microarray analysis, *Mutat. Res.*, 597, 98–112, 2006.
23. Wolff, S., Aspects of the adaptive response to very low doses of radiation and other agents, *Mutat. Res.*, 358, 135–42, 1996.
24. Fry, M., Deterministic effects, *Health Phys.*, 80, 338–43, 2001.
25. Hall, E., and Giaccia, A., *Radiobiology for the Radiologist*, Philadelphia: Lippincott Williams & Wilkins, 2005.
26. Thomas, O., Mahe, M., Campion, L., Bourdin, S., Milpied, N., Brunet, G., Lisbona, A., Le, M. A., Moreau, P., Harousseau, J., and Cuilliere, J., Long-term complications of total body irradiation in adults, *Int. J. Radiat. Oncol. Biol. Phys.*, 49, 125–31, 2001.
27. Pierce, D. A., Shimizu, Y., Preston, D. L., Vaeth, M., and Mabuchi, K., Studies of the mortality of atomic bomb survivors. Report 12, Part I. Cancer: 1950–1990, *Radiat. Res.*, 146, 1–27, 1996.
28. Shimizu, Y., Pierce, D. A., Preston, D. L., and Mabuchi, K., Studies of the mortality of atomic bomb survivors, Report 12, Part II, Noncancer mortality: 1950–1990, *Radiat. Res.*, 152, 374–89, 1999.
29. Neel, J. V., Schull, W. J., Awa, A. A., Satoh, C., Kato, H., Otake, M., and Yoshimoto, Y., The children of parents exposed to atomic bombs: Estimates of the genetic doubling dose of radiation for humans, *Am. J. Hum. Genet.*, 46, 1053–72, 1990.
30. International Commission on Radiological Protection, *Radiological Protection and Safety and Pregnancy*, ICRP Report 77, Oxford: Pergamon Press, 1999.
31. International Commission on Radiological Protection, *Biological Effects after Prenatal Irradiation (Embryo and Fetus)*, ICRP Publication 90, Oxford: Pergamon Press, 2003.
32. Doll, R., and Wakeford, R., Risk of childhood cancer from fetal irradiation, *Brit. J. Radiol.*, 70, 130–39, 1997.
33. Hanahan, D., and Weinberg, R., The hallmarks of cancer, *Cell*, 100, 57–70, 2000.
34. Okada, H., and Mak, T. W., Pathways of apoptotic and non-apoptotic death in tumour cells, *Nat. Rev. Cancer*, 4, 592–603, 2004.
35. International Commission on Radiological Protection, *Low-Dose Extrapolation of Radiation Related Cancer Risk*, ICRP Publication 99, Oxford: Pergamon Press, 2006.
36. Harris, C. C., Structure and function of the p53 tumor suppressor gene: Clues for rational cancer therapeutic strategies, *J. Natl. Cancer Inst.*, 88, 1442–55, 1996.

37. Bonner, J. A., Harari, P. M., Giralt, J., Azarnia, N., Shin, D. M., Cohen, R. B., Jones, C. U., Sur, R., Raben, D., Jassem, J., Ove, R., Kies, M. S., Baselga, J., Youssoufian, H., Amellal, N., Rowinsky, E. K., and Ang, K. K., Radiotherapy plus cetuximab for squamous-cell carcinoma of the head and neck, *N. Engl. J. Med.*, 354, 567–78, 2006.

38. Hayflick, L., Mortality and immortality at the cellular level: A review, *Biochemistry (Mosc.)*, 62, 1180–90, 1997.

39. Folkman, J., Role of angiogenesis in tumor growth and metastasis, *Semin. Oncol.*, 29, 15–18, 2002.

40. Jain, R., Normalization of tumor vasculature: An emerging concept in antiangiogenic therapy, *Science*, 307, 58–62, 2005.

41. Folkman, J., Antiangiogenesis in cancer therapy: Endostatin and its mechanisms of action, *Exp. Cell Res.*, 312, 594–607, 2006.

42. Sporn, M. B., The war on cancer, *Lancet*, 347, 1377–81, 1996.

43. Trosko, J., Role of low-level ionizing radiation in multi-step carcinogenic process, *Health Phys.*, 812–22, 1996.

44. International Commission on Radiological Protection, *Genetic Susceptibility to Cancer*, ICRP Publication 79, Oxford: Pergamon Press, 1999.

45. Cardis, E., Richardson, D., and Kesminiene, A., Radiation risk estimates in the beginning of the 21st century, *Health Phys.*, 80, 349–61, 2001.

46. Brenner, D. J., Doll, R., Goodhead, D. T., Hall, E. J., Land, C. E., Little, J. B., Lubin, J. H., Preston, D. L., Preston, R. J., Puskin, J. S., Ron, E., Sachs, R. K., Samet, J. M., Setlow, R. B., and Zaider, M., Cancer risks attributable to low doses of ionizing radiation: Assessing what we really know, *Proc. Natl. Acad. Sci. U.S.A.*, 100, 13761–66, 2003.

47. Parfit, T., Opinion remains divided over Chernobyl's true toll, *Lancet*, 367, 1305–6, 2006.

48. Carnes, B., Groer, P., and Kotek, T., Radium dial workers: Issues concerning dose response and monitoring, *Radiat. Res.*, 147, 707–14, 1997.

49. Cardis, E., Howe, G., Ron, E., Bebeshko, V., Bogdanova, T., Bouville, A., Carr, Z., Chumak, V., Davis, S., Demidchik, Y., Drozdovitch, V., Gentner, N., Gudzenko, N., Hatch, M., Ivanov, V., Jacob, P., Kapitonova, E., Kenigsberg, Y., Kesminiene, A., Kopecky, K. J., Kryuchkov, V., Loos, A., Pinchera, A., Reiners, C., Repacholi, M., Shibata, Y., Shore, R. E., Thomas, G., Tirmarche, M., Yamashita, S., and Zvonova, I., Cancer consequences of the Chernobyl accident: 20 years on, *J. Radiol. Prot.*, 26, 127–40, 2006.

50. Hall, E. J., Intensity-modulated radiation therapy, protons, and the risk of second cancers, *Int. J. Radiat. Oncol. Biol. Phys.*, 65, 1–7, 2006.

51. Nambi, K., and Soman, S., Further observations on environmental radiation and cancer in India, *Health Phys.*, 59, 339–44, 1990.

52. Cohen, B. L., Test of the linear no-threshold theory of radiation carcinogenesis for inhaled radon decay product, *Health Phys.*, 68, 157–74, 1995.

53. National Council on Radiation Protection and Measurements, *Risk Estimates for Radiation Protection*, NCRP Report 115, Bethesda, MD: NCRP, 1993.

54. Wigg, D., *Applied Radiobiology and Bioeffect Planning*, Madison, WI: Medical Physics Publishing, 2001.

55. Fowler, J., The linear-quadratic formula and progress in fractionated radiotherapy, *Brit. J. Radiol.*, 62, 679–94, 1989.

56. Dale, R., The application of the linear quadratic dose effect equation to fractionated and protracted radiotherapy, *Brit. J. Radiol.*, 58, 515–28, 1985.

57. Sinclair, W. K., The linear no-threshold response: Why not linearity? *Med. Phys.*, 25, 285–90, 1998.
58. Clarke, R., 21st century challenges in radiation protection and shielding: Draft 2005 recommendations of ICRP, *Radiat. Prot. Dosimetry*, 115, 10–15, 2005.
59. Clarke, R., and Valentin, J., A history of the international commission on radiological protection, *Health Phys.*, 88, 717–32, 2005.
60. Strom, D., Cameron, J., Cohen, B., Mossman, K., Nussbaum, R., Sinclair, W., and Webster, E., Controversial issues: The linear no-threshold (LNT) debate, *Med. Phys.*, 28, 273–300, 1998.
61. Charles, M. W., LNT: An apparent rather than a real controversy? *J. Radiol. Prot.*, 26, 325–29, 2006.
62. Tubiana, M., Aurengo, A., Averbeck, D., and Masse, R., The debate on the use of linear no threshold for assessing the effects of low doses, *J. Radiol. Prot.*, 26, 317–24, 2006.
63. Chen, W., Luan, Y., Shieh, M., et al., Is chronic radiation an effective prophylaxis against cancer? *J. Am. Physicians Surgeons*, 9, 6–10, 2004.
64. Land, C. E., Tokunaga, M., Tokuoka, S., and Nakamura, N., Early-onset breast cancer in A-bomb survivors, *Lancet*, 342, 237, 1993.
65. Pukkala, E., Auvinen, A., and Wahlberg, G., Incidence of cancer among Finnish airline cabin attendants, 1967–92, *BMJ*, 311, 649–52, 1995.
66. Miller, A. B., Howe, G. R., Sherman, G. J., Lindsay, J. P., Yaffe, M. J., Dinner, P. J., Risch, H. A., and Preston, D. L., Mortality from breast cancer after irradiation during fluoroscopic examinations in patients being treated for tuberculosis, *N. Engl. J. Med.*, 321, 1285–89, 1989.
67. Hoffman, D. A., Lonstein, J. E., Morin, M. M., Visscher, W., Harris, B. S., III, and Boice, J. D., Jr., Breast cancer in women with scoliosis exposed to multiple diagnostic x rays, *J. Natl. Cancer Inst.*, 81, 1307–12, 1989.
68. Lundell, M., Mattson, A., Karlsson, P., Holmberg, E., Gustafsson, A., and Holm, L. E., Breast cancer risk after radiotherapy in infancy: A pooled analysis of two Swedish cohorts of 17,202 infants, *Radiat. Res.*, 151, 626–32, 1999.
69. Clarke, M., Collins, R., Darby, S., Davies, C., Elphinstone, P., Evans, E., Godwin, J., Gray, R., Hicks, C., James, S., MacKinnon, E., McGale, P., McHugh, T., Peto, R., Taylor, C., and Wang, Y., Effects of radiotherapy and of differences in the extent of surgery for early breast cancer on local recurrence and 15-year survival: An overview of the randomised trials, *Lancet*, 366, 2087–106, 2005.
70. Carmichael, A., Sami, A. S., and Dixon, J. M., Breast cancer risk among the survivors of atomic bomb and patients exposed to therapeutic ionising radiation, *Eur. J. Surg. Oncol.*, 29, 475–79, 2003.
71. Travis, L. B., Hill, D. A., Dores, G. M., Gospodarowicz, M., van Leeuwen, F. E., Holowaty, E., Glimelius, B., Andersson, M., Wiklund, T., Lynch, C. F., Van't Veer, M. B., Glimelius, I., Storm, H., Pukkala, E., Stovall, M., Curtis, R., Boice, J. D., Jr., and Gilbert, E., Breast cancer following radiotherapy and chemotherapy among young women with Hodgkin disease, *JAMA*, 290, 465–75, 2003.
72. Bhatia, S., Robison, L. L., Oberlin, O., Greenberg, M., Bunin, G., Fossati-Bellani, F., and Meadows, A. T., Breast cancer and other second neoplasms after childhood Hodgkin's disease, *N. Engl. J. Med.*, 334, pp. 745–751, 1996.

4

Radiation Detection and Simulation Methods

Jamie V. Trapp and Peter Johnston

CONTENTS

Chapter 2 provided an overview of radiation physics for radiation protection. The principles of radiation interactions with matter can be utilized to measure radiation with detectors or to accurately predict many aspects of radiation via Monte Carlo simulations. This chapter will provide an overview of such methods and the principles of their operation. Thorough treatment of radiation detection and simulations would require several texts; therefore, if more comprehensive information is required, the reader is directed to specific texts on radiation such as *Radiation Detection and Measurement* by G. F. Knoll, *Medical Radiation Detectors: Fundamental and Applied Aspects* by N. F. Kember, and so forth.

4.1   Radiation Detection Methods

When designing or selecting a radiation detector for a particular application, care should be taken to ensure that the most appropriate detector is used. The detector should be sensitive to the type of radiation of interest and be sensitive to the range of radiation measurements undertaken. In particular, it is important to realize that uncharged particles cannot be measured directly and are invariably detected by the secondary charged particles they generate; that is, secondary electrons for photon detection and recoil protons are commonly used for fast-neutron detection.

There are many other considerations when choosing a detector that may include whether the user wishes to count individual radiation events, whether spectroscopy is performed, and whether there is a requirement for energy dependence, count rate dependence, directional dependence, and so on.

4.1.1   Gas-Filled Detectors

The method of operation of gas-filled detectors is that radiation causes ionization of gas contained between two electrodes in a chamber, and the resulting electric charge is collected as a current or pulse. These detectors are most commonly used for counting.

When a photon moves through a gas it can interact with the atoms within the gas via photoelectric effect producing a photoelectron or Compton scattering producing a Compton electron, or create an electron-positron pair through pair production. Likewise, alpha and beta radiation cause the ionization of the gas particles during their energy loss processes. The average energy lost per ionization event (termed the W value) is small compared to most incident radiation energies, and so many ionizations can occur from a radiation quantum. The result is that the number of ion pairs created per quantum is very nearly proportional to its initial energy. For example, the W value for fast electrons in air is 33.8 eV/ion pair, and so an electron with 2 MeV of kinetic energy could be expected to produce almost sixty thousand ion pairs before all of its energy is lost.

To collect the ion pairs, or more specifically the ejected electrons, an electric field is established between electrodes in the gas chamber (effectively turning the arrangement into a capacitor). The geometry is normally in the form of either two parallel plates or a cylinder with a central anode (the wall of the cylinder either being a cathode or kept at ground potential), as shown in Figure 4.1. As electrons are removed from the atoms within the gas, they move under influence of the electric field, which exists due to the different potentials of the electrodes. Electrons are collected by the anode and as an electric current, which is measured by an *electrometer*.

The operation of gas-filled detectors varies according to the magnitude of the voltage applied between the electrodes. Figure 4.2 shows the approximate trend of the collection of ions with increasing voltage, and these char-

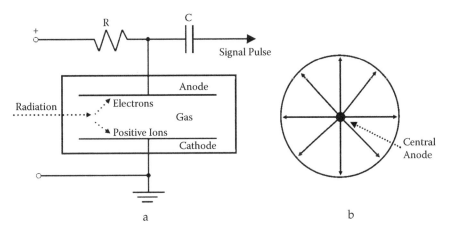

FIGURE 4.1
Different geometries of gas-filled detectors: parallel plate and cylindrical. Radiation causes ionization of the gas in the chamber and electrons are collected on the anode. Note the greater density of electric field lines near the anode in cylindrical geometry.

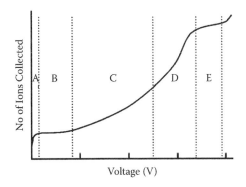

FIGURE 4.2
The number of ions collected by a gas-filled detector as the voltage between the electrodes is varied. Explanation of the regions is given in the text.

acteristics are used to design ionization chambers, proportional counters, or Geiger-Muller counters. Region A of Figure 4.2 is the recombination region, where although ionization of the gas occurs, the electric field between the electrodes is not sufficient to accelerate all of the electrons enough to reach the anode before undergoing recombination with other atoms within the gas (i.e., they are "captured" back into an atom). Increasing the voltage will enable the chamber to operate as an ionization chamber in Region B. At the voltages in this region there is sufficient electric field to ensure that almost all electrons undergo sufficient acceleration to reach the anode before recombination occurs. Moderate increases in voltage in this region will not result in much increase in collection efficiency, as most electrons are collected anyway. As the voltage increases to Region C, the detector enters the proportional counter region. At these voltages the electrons gain sufficient kinetic energy to cause ionization of other atoms in the gas before reaching the anode (called a *Townsend avalanche*); hence there is a multiplication of the signal, typically up to 10^4 or 10^5 times. To produce sufficient electric field for this effect, a cylindrical design must be used to ensure a sufficiently high electric field near the anode (note the density of electric field lines near the anode in Figure 4.1b). Region D is of limited proportionality, which is not generally used in radiation detection. Region E is the Geiger-Muller region, where gas multiplication has reached a saturation level of around 10^8 and the discharge occurs along the entire length of the anode wire. Operating at this region is more efficient in detection of radiation than proportional counters; however, there is no energy discrimination due to the saturated discharge. Voltages larger than that of the Geiger-Muller region are generally not used and can damage equipment. Although the above discussion describes the effect of increasing voltage of a single detector, in practice commercially available detectors are generally designed for operation at a specific voltage.

4.1.2   Scintillation Detectors

Scintillation detectors operate on the principle that certain materials produce light in response to radiation (i.e., they scintillate). Scintillation light is produced by the de-excitation of either the crystal lattice or individual molecules following an initial excitation due to radiation.

These detectors can be in the form of a crystal, plastic, liquid, or even a gas, and the detection material usually has a greater density than that of gas-filled detectors, and so there is a greater probability that radiation will interact within the detector leading to greater efficiency than gas-filled detectors. Scintillating detectors have moderate energy resolution and fast response times and are therefore suitable for spectroscopy.

Scintillating detectors are broadly classified as either inorganic or organic. Inorganic scintillators are structured crystals and emit light as a result of excitation of the crystal lattice. The electrons within the crystal occupy either the valence band, where they are bound to atoms or locations within the crystal, or the conduction band, where they are free to migrate throughout the crystal (Figure 4.3). The energy of the conduction band is higher than that of the valence band. A certain amount of energy is required to raise an electron from the valance band to the conduction band. The return of the electron from the conduction band to the valence band requires a release of energy, which in this case is in the form of a scintillation photon. The gap in energy between the bands can sometimes be reduced by the introduction of an impurity, as shown on the right-hand side of Figure 4.3. Decreasing the band gap improves efficiency of an inorganic scintillator and subsequently

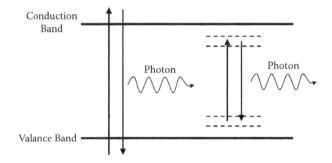

FIGURE 4.3
Energy states of an inorganic scintillating crystal. An incoming radiation raises an electron from the valence band to the conduction band (*left*). As the electron de-excites a photon carries the excess energy away. Doping the crystal can reduce the energy gap between the bands, making the crystal more sensitive (*right*).

increases the wavelength of the scintillation light (e.g., a sodium iodide detector scintillates at 303 nm, but doping with thallium increases the scintillation light to 410 nm).

Organic scintillators emit light as a result of energy-level transitions of single molecules. The advantage of this is that the scintillator does not need to be in the form of a crystal and therefore can be manufactured as a solid, liquid, moldable plastic, film, and so forth. As a charged particle moves through the scintillating material it loses energy to nearby molecules. This energy is absorbed by the molecules and raises the electron configuration to an excited state, as shown in Figure 4.4. The excited states can be either an excited electron state of several electronvolts or an excited vibrational energy of a fraction of an electronvolt. The molecule loses its vibrational energy through thermal processes, and the energy from the excited electron state is lost through emission of a photon.

The scintillation light emitted during the radiation event must be converted into an electrical signal for measurement. This is most commonly achieved through the use of a photomultiplier tube, diagrammatically represented in Figure 4.5. The photomultiplier tube consists of a photocathode and a number of dynodes. In response to a scintillation photon the photocathode emits a small number of electrons. These electrons are attracted to a nearby dynode, which is held at a positive electrical potential. Collision of each electron with the dynode causes the ejection of between about fifty and a hundred new electrons from the dynode's surface. Some of these electrons (say, three to five) escape the positive field of the dynode and are attracted to another nearby dynode held at yet a higher potential. The process is continued through a series of dynodes held at increasingly higher potential until at the last stage they are collected as an electrical pulse. The photomultiplier tube is an efficient means of both converting light to electric current and

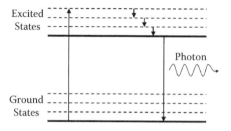

FIGURE 4.4

Electronic states of an organic scintillating molecule. Energy absorbed from a passing radiation excites the molecule. As the molecule de-excites the excess energy is carried away by a photon.

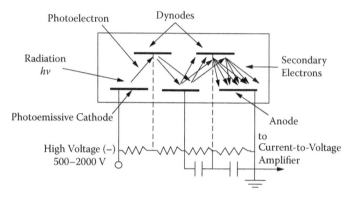

FIGURE 4.5

Diagrammatic representation of a photomultiplier tube.

amplifying the signal (e.g., if the multiplication is five times at each dynode, and there are ten dynodes in the series, the final multiplication will be 5^{10}).

Both of the scintillating detectors described above only absorb a relatively small amount of energy from charged particles with each interaction. This means that a single incident radiation can interact several thousands of times before the energy is lost, resulting in several thousand scintillation photons. The number of scintillation events will vary according to the energy of the incident radiation; therefore, by measuring the intensity of scintillation photons the energy of the incident radiation can be determined and thus spectroscopy performed.

One further consideration when using scintillators as detectors is the presence of Cerenkov light, which can occur when measuring high-energy charged particles. This radiation is caused when a charged particle transits a medium, in this case the detector, at a speed exceeding the speed of light in that medium (note that this is less than the speed of light in a vacuum). As the particle transits the medium it excites the local electromagnetic fields, and as the atoms return to their ground state they de-excite through the emission of photons. Destructive interference of the emitted photons normally occurs; however, if the transiting particle is traveling faster than the emitted photons can travel, then constructive interference will occur, seen as Cerenkov light. Care must be therefore be taken when using scintillators to measure high-energy charged particles, and strategies such as parallel light guides have been employed to reduce these effects.[1]

4.1.3   Semiconductor Detectors

Semiconductor detectors are formed of crystals that respond to radiation energy deposition by raising an electron from the valence band to the conduction band. These electrons are collected by applying an electric field across the detector crystal, causing them to be attracted to the positive electrode. The raising of electrons to the conduction band results in electron vacancies, or holes, which are attracted to the negative electrode. Semiconductor detectors are operated with biases of up to several thousand volts. To reduce thermal noise, some detectors are cooled by liquid nitrogen, providing improved resolution.

In semiconductor detectors the energy gap between the conduction and valence bands is considerably smaller than that in scintillation detectors, and so the number of electrons mobilized per unit energy is much greater, resulting in more effective energy discrimination. The resolution can be further improved by the addition of dopants to reduce the band gap. However, doping the crystal can come at a price; in the case of Ge detectors, which can be stored at room temperature between use, the addition of Li to form a Ge(Li) detector requires permanent storage at liquid nitrogen temperatures to prevent the migration of the Li throughout the detector.

4.1.4   Thermoluminescent Dosimeters

Imperfect or doped crystal structures contain electron traps, whereby electrons previously raised to the conduction band do not always fully return to the valence band (see Figure 4.6). These traps occur in the location of lattice imperfections, or at impurity atom sites within the crystal, and can "hold" the electron for a long time, in many cases years. Thermoluminescent dosimeters (TLDs) utilize this phenomenon as a means of storing the radiation energy deposited in the crystal as a method of radiation detection.

When the crystal is heated the electrons are released from the traps and return to the valence band, and the energy released is emitted in the form of light photons. The intensity of light released is proportional to the radiation dose absorbed in the crystal. The light output from TLDs is measured using specialized equipment that simultaneously heats the crystals and measures the light. TLDs contain many types of trap at different interband energies so that some require more heat than others to release the electrons; therefore, light is emitted over a range of temperatures up to several hundred degrees Celsius.

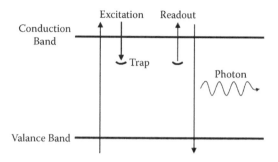

FIGURE 4.6

Energy process of thermoluminescent dosimeters. An incident radiation causes an electron to be raised to the conduction band; however, it does not de-excite fully and is caught in a trap. The application of heat will raise the electron out of the trap to the conduction band and de-excitation can then occur.

In medical dosimetry the most common material used in TLDs is LiF, which has the advantage of being nearly tissue equivalent (effective atomic number of 8.4 compared to tissue, whose number is 7.2) and therefore has a reasonably flat energy response. The disadvantage for radiation protection purposes is that LiF is not sensitive enough for use in measuring monthly exposures for workers, so $CaSO_4$:Dy is used for TLDs in personal monitoring.

4.1.5   Film

Film dosimeters incorporate photographic emulsion of (typically) silver halide crystals and gelatin, which is coated onto a transparent polyester base. Incident radiation raises the electrons in the crystals from the valence band to the conduction band, and so they are free to migrate throughout the crystal. Some electrons recombine with electron holes; however, some are trapped in electron traps in the same manner as that described for TLDs. Around the region of the electron traps a slightly negative electric field forms due to the accumulation of the electrons' charge, and this electric field attracts small amounts of positively charged mobile interstitial silver ions. These ions combine with the trapped electrons and reduce to metallic silver. The combination of all of these metallic silver atoms throughout the film is said to form a *latent image*. Unless the film is overexposed or saturated, only a fraction of the grains interact with the incident radiation, and so the film can be developed by chemically removing a portion of the unexposed grains. Consequently, the optical density of a portion of film relates to the radiation dose received.

Although film can be used for medical imaging, the sensitivity of the film to radiation increases with the size of the crystals, and so film used for personal monitoring does not have sufficient spatial resolution for good-quality medical images. Likewise, film used for imaging and linear accelerator testing is generally too insensitive for the small doses received for personal monitoring purposes.

4.1.6 Chemical and Gel Dosimeters

Some radiation detectors rely on chemical changes in substances resulting from ionizations. Fricke dosimeters[2] are an aqueous solution of ferrous ions (Fe^{2+}). Ionizing radiation causes a number of reactions in the water base of the solutions, which leads to the transformation of ferrous ions to ferric (Fe^{3+}) ions. The number of ions transformed is related to the radiation dose absorbed in the solution; therefore, by measurement of the relative quantity of ferric ions in a Fricke solution, one can determine the radiation dose that the solution has received. Other chemical dosimeters include those based on polymerization reactions such as that observed in crystalline acrylamide,[3] or colorization of dyes.[4]

An innovative approach of using chemical dosimeters has been their infusion within a gel or solid to provide a spatial map of radiation dose. This occurred at least as early as 1950;[4] however, it was not until 1984 that a reliable method of spatially measuring the chemical changes with magnetic resonance imaging (MRI) was suggested.[5] The postirradiation dose map in gel and solid dosimeters often undergo a number of measurable changes that can be utilized to extract a dose distribution through techniques such as MRI, x-ray computed tomography (CT), optical CT, ultrasound, raman spectroscopy, and so on.

4.1.7 Neutron Detection Methods

Because neutrons have no net electrical charge and are not ionizing, the detection methods outlined above are not suitable for their detection. Neutrons are not detected through direct measurement, but rather through the measurement of charged particles that are released in response to interaction of neutrons with specific elements. Neutron detection methods are broadly separated into detection of thermal neutrons and detection of fast neutrons.

Thermal neutron detection is the measurement of the reaction products of thermal neutrons with ^{10}B, ^{6}Li, or ^{3}He. Thermal neutron capture by ^{10}B has a high cross section for reaction and results in ^{7}Li, an alpha particle, and either 2.31 MeV (96%) or 2.79 MeV (4%) of energy. Boron-based neutron detectors are generally configured as a proportional counter filled with a boron-containing gas, normally BF_3. The energy released in the reaction results in a relatively large pulse height, which can be used to discriminate the neutron interaction from associated gamma reactions. The *lithium (^{6}Li) reaction* results in a tritium nucleus, an alpha particle, and 4.78 MeV of energy. Detectors utilizing this reaction are manufactured as crystals and configured as scintillators. The advantage of this type of detector is that the larger Q value of the reaction allows for better discrimination between the neutron reaction and gamma rays, but the reaction cross section is much lower than that of boron, and so there is a loss of efficiency. The *helium (^{3}He) reaction* results in a tritium nucleus, a hydrogen nucleus, and 0.76 MeV of energy, and detectors are configured as proportional counters at high pressure. There is a high-interaction cross section, and therefore a high efficiency; however, the smaller Q value, the poorer the discrimination between neutrons and gamma radiations.

The interaction cross sections of the above materials are generally too low for fast neutrons to allow efficient counting. Fast-neutron detectors operate either by slowing the neutrons down to thermal energies prior to detection by use of a moderator or through detection based on other reactions. In the first method, the detector is based on one of the designs used to measure thermal neutrons; however, it is surrounded by a thickness of hydrogen-rich moderating material such as paraffin. Increasing the thickness of the moderator on one hand improves efficiency by slowing neutrons down further through a longer path length through the material, but on the other hand decreases efficiency if the moderator is too thick, as some neutrons will be completely absorbed by the moderator or deflected away before reaching the detector. Therefore, the optimal moderator thickness for a neutron detector varies according to the energy of the neutrons. For a detector with a fixed moderator thickness the efficiency of the detector will have an energy dependence. Other neutron detectors do not rely on moderation but on reactions directly induced by fast-neutron interactions, which can measure the energy of the interaction products and be used for spectroscopy.

Recently another type of dosimeter for neutrons has become commercially available: superheated drop or bubble dosimetry.[6,7] These detectors employ the well-established concept of the bubble chamber for particle detection by using microscopic (diameter about 100 µm) droplets of superheated liquid in a polymer gel matrix. Neutrons traversing a microbubble deposit sufficient energy to evaporate the superheated droplet that forms a visible bubble. The number of the bubbles is related to the neutron dose. Advantages of bubble dosimeters are the high specificity for neutrons in a mixed gamma neutron field and the fact that they can be reused by pressurizing the gel. By changing the number of the microbubbles in the gel, the sensitivity can be varied. Bubble dosimeters have been used for neutron dosimetry in linac bunkers, and with the size of a test tube they are also small enough to be used as a personal dosimeter.

4.2 Radiation Simulations through Monte Carlo Modeling

Monte Carlo modeling of radiation transport is the gold standard in the prediction of a radiation field. The method uses computer simulations to model the paths and interactions of individual primary and secondary radiations from some initiating event through one or more media (e.g., simulated body tissues, shielding materials, detector materials, etc.). The simulation runs until all of the initial particles are annihilated or fall below a threshold parameter where their behavior is no longer of interest, for example, a particle's energy falling below a cutoff limit, a particle moving outside of a region of interest, and so forth. The cascade of particles and where they deposit energy are often described as a history. Whenever the particles interact, the properties of the particles can be monitored to score physical parameters of interest, for example, dose, KERMA.

There are many Monte Carlo codes for radiation transport commonly in use, for example, GEANT, MCNP, and EGSnrc. The various codes reflect the interests of the developers. GEANT is a code originally developed for high-energy physics applications, typically in particle physics. MCNP is widely used for problems involving neutrons, for example, the reactivity of reactor cores. EGSnrc is named for Electron Gamma Shower (EGS), which is the most commonly used code in medical physics applications. EGSnrc has many usercodes developed for it that may be of use in medical physics, for example, BEAMnrc, which is designed for modeling problems associated with megaelectronvolt electron linear accelerators used in radiotherapy.

In each case, there are common elements of the code, including the geometry of the system to be modeled, the materials in each region of the model, a system for scoring the results of each interaction, and the physics behind the interactions between modeled particles and the materials, that is, the radiation interaction parameters of the incident particles in those materials. Random variables from distributions consistent with the physics are generated to create these histories. Models are typically run for a given number of initiating events or histories or until a statistical threshold is satisfied.

Scoring of the outputs may be quite complex; for example, for the modeling of pulse height spectra, the important parameter is the total energy deposited in the detector in a short period of time. This usually corresponds to all the energy deposited in the detector from a history generated by an incident radiation or emanating from a radioactive decay site. Scoring can also be used to generate dose, KERMA, or even allow the problem to be broken into a sequence of parts using *phase-space* files. This is necessary in problems that are so inefficient that the entire problem may never yield an adequate statistical result, but consideration of subproblems each of high efficiency may provide a successful solution.

In medical applications, typically dose or dose to a specific tissue are the parameters of interest. Usually these problems involve photons and electrons, and so the physics built into EGSnrc and its usercodes is sufficient for many problems. The EGSnrc system is very powerful, allowing the modeling of general problems involving electrons (both negatrons and positrons) and photons in complex geometrical arrangements. This involves the direct coding of the problem in the MORTRAN language (a preprocessor for FORTRAN), including the geometry of the situation to be modeled and the definition of routines indicating the distance to the nearest region boundary (HOWNEAR) and the distance along the current track to a boundary (HOW-FAR). It is often possible to model a problem in cylindrical or rectilinear coordinates using standard usercodes DOSRZnrc and DOSXYZnrc. For the modeling of clinical linear accelerators used in radiotherapy, the BEAMnrc code has been developed with geometrical objects designed for describing the components of such a device. EGSnrc usercodes allow an input file to be defined to describe the model without the need of direct coding of the problem in the MORTRAN language, and are therefore much more suitable for rapid model development. The disadvantage of such usercodes is that they are less efficient in many cases because the usercode has to deal with the generalized problem rather than a special case.

BEAMnrc is typically used together with DOSXYZnrc. The output of BEAMnrc is typically a *phase-space* file containing a list of the energies and directions of radiations passing through a plane corresponding to the exit of a beam from a linear accelerator. This phase-space file may then be used as the radiation source for a subsequent model, typically using DOSXYZnrc, which describes a target for the beam that might be rectilinear or even a voxelated three-dimensional structure derived from a CT scan. This is an example where the inefficient process of generating the radiation from the accelerator exit need only be performed once per setup of accelerator jaws and used for many subsequent models.

References

1. Beddar, A. S., T. R. Mackie, and F. H. Attix, Water-equivalent plastic scintillation detectors for high-energy beam dosimetry. I. Physical characteristics and theoretical considerations, *Phys. Med. Biol.*, 1992, 37: 1883–1900.
2. Fricke, H., and S. Morse, The chemical action of roentgen rays on dilute ferrous sulfate solutions as a measure of radiation dose, *Am. J. Roentgenol. Radium Therapy Nucl. Med.*, 1927, 18: 430–32.
3. Mesrobian, R. B., P. Ander, D. S. Ballantine, and G. J. Dienes, Gamma-ray polymerization of acrylamide in the solid state, *J. Chem. Phys.*, 1954, 22: 565–66.
4. Day, M. J., and G. Stein, Chemical effects of ionizing radiation in some gels, *Nature*, 1950, 166: 1146–47.
5. Gore, J. C., Y. S. Kang, and R. J. Schulz, Measurement of radiation dose distributions by nuclear magnetic resonance (NMR) imaging, *Phys. Med. Biol.*, 1984, 29: 1189–97.
6. Harper, M., and M. Nelson, Experimental verification of a superheated liquid droplet (bubble) neutron detector theoretical model, *Rad. Prot. Dosim.*, 1993, 47: 535–42.
7. Nelson, M., and R. Gordon, Comparison of neutron measurements at linacs using bubble dosimeters to other neutron detectors, *Rad. Prot. Dosim.*, 1993, 47: 547–50.

5

Managing Radiation in the Workplace

Peter Johnston, Jamie V. Trapp, and Tomas Kron

CONTENTS

Radiation is a hazard to people, animals, and the environment, affecting both everyday life and the workplace. Therefore, it is important that adequate protection measures are taken. As a general rule, radiation protection measures are easier to apply in the controlled environment of a workplace. This chapter aims to highlight some important aspects of radiation protection with particular reference to the medical environment.

An international system of radiological protection has been developed to protect human beings and the environment. It is important to recognize that this system of protection fits within the overall context of a hierarchy of safety controls that apply to all hazards. This hierarchy provides controls in the following order:

1. If possible, the hazard should be eliminated.
2. The safer alternative should be substituted if available.
3. The hazard should be isolated.
4. Engineering controls should limit the impact of the hazard.
5. Administrative controls should be employed to limit the hazard.
6. Personal protective equipment (PPE) should be employed in dealing with the hazard.

These guiding principles in minimizing the hazard are central to the system of radiological protection that is internationally acknowledged. As in other areas of occupational health and safety, the maintenance of an active safety culture and the involvement of senior management in safety are vital in radiation protection.

The overall safety program for an enterprise should include personal protection and monitoring, as well as a documented radiation management plan, system procedures, standard operating procedures, and protocols. The radiation management plan should promote a good safety culture in the use of ionizing radiation and be reviewed and updated regularly to maintain currency in terms of national and international standards. The plan should include:

- Methods for compliance with regulations and standards
- Optimization of radiation exposures and the development of appropriate dose constraints
- Requirements for approvals
- Responsibilities of staff
- Descriptions of the means of controlling exposures
- Designation of controlled and supervised areas
- Authorizations and delegations of authority to individuals
- Training, induction, and accreditation as required
- Procedures for monitoring personnel, equipment, and the environment
- Record keeping requirements
- Procedures for dealing with incidents, accidents, and emergencies

5.1   Exposure Categories in the Recommendations of the ICRP

Regulation of radiation safety throughout the world is currently based on the *1990 Recommendations of the ICRP*;[1] however, from 2007 the International Commission on Radiological Protection (ICRP) has adopted new recommendations.[2,3] The new recommendations are intended to be an evolutionary progression of the system of radiation protection and will take years to be promulgated through to national regulation. The following description is based on both the 1990 and 2007 recommendations.

The 1990 recommendations describe exposures in terms of practices and interventions, which are redefined in the 2007 recommendations to be planned situations, existing situations, and emergency situations. The following discussion relates to practices in the sense of the 1990 recommendations (i.e., applicable to planned situations in the newer framework).

The 1990 recommendations describe exposures in three categories: occupational, medical, and public (also see Chapter 1). For the medical workplace all three types of exposure are relevant and must be considered. **Occupational exposure** is incurred at work by an individual knowingly working with ionizing radiation. The exposure is a result of the nature of the work and includes most, but not all, sources of exposure incurred at the workplace, that is, only those that are "the responsibility of the operating management." This includes practices where exposure is increased simply by position, for example, doses to aircrew are included in occupational exposure. The occupationally exposed persons are typically monitored for exposure to ionizing radiation and have regular training.

Public exposure is the exposure of any member of the general public. Public exposures cover all those arising from a particular activity involving the use of radiation. If an exposure is not occupational or medical, it is considered to be public.

Medical exposure is the exposure of patients as part of diagnostic procedures or as part of their treatment, either as the patient or as a supporter of the patient during the medical procedure, but does not include those persons who are occupationally exposed as part of the procedure. Medical exposures also include volunteers in medical research. Only in the latter case is it possible to apply meaningful dose constraints—in the case of patients the prescription provides justification, and it is always assumed that the potential benefits outweigh the risks for the individual patient.

5.2 Control of Exposures

The six points listed at the beginning of the chapter to indicate the hierarchy of safety controls can be considered from three different angles: administrative, instrumental/engineering, and ongoing quality assurance. The following sections address these aspects.

TABLE 5.1

Dose Limits Recommended in ICRP60

	Occupational Exposure	Public Exposure	Medical Exposure
Effective dose (1 year)	50 mSv	1 mSv	No dose limit applicable
Effective dose (5 consecutive years)	100 mSv	5 mSv	
Dose to the eye	150 mSv per year	15 mSv per year	
Skin dose	500 mSv per year	50 mSv per year	
Unborn child	1 mSv for lifetime of pregnancy		

5.2.1   Administrative Considerations

The control of public exposures is the most complex one given that administrative controls are difficult to implement. Similarly, the use of personal protective devices and monitoring of exposure to individuals is not usually appropriate. Given that the practice is justified (i.e., the hazard cannot be eliminated or substituted), then the hazard must be isolated or engineering controls implemented to protect the public. Dose limits for protection of the public are imposed by regulatory authorities, based on ICRP recommendations. The recommended annual limits are 1 mSv effective dose, 15 mSv to the lens of the eye, and 50 mSv to the skin. Table 5.1 lists the dose limits provided in ICRP Report 60.[1]

Such limits apply to doses from all practices. ICRP recommends that a constrained optimization approach be used applying a dose constraint that is selected to allow for significant contributions from other sources. This is usually implemented to provide a dose constraint for the critical group of one-third of the dose limit from a single source. In the context of a hospital this value makes sense when considering that a location may be in proximity to a radiotherapy department as well as a nuclear medicine and diagnostic radiology center. For example, if an office space is between a nuclear medicine center and a radiotherapy department the total dose limit to the office workers is 1 mSv per year under ICRP60; however, the dose constraint is such that each department must be designed so that it contributes no more than 0.3 mSv per year to the office workers. The dose limit applies to the individual receiving dose from all sources under normal conditions.[3] The constraints provide for the control of the individual sources.

Medical exposures occur in a much more controlled environment. In relation to medical exposure of the patient, the doses may be very high, especially in the case of radiation therapy. Nevertheless, optimization of the practice means that doses received to untargeted tissues of the patient should be kept as low as reasonably achievable (the ALARA principle) without compromising the therapeutic outcome. This criterion is gaining in importance as outcomes for patients improve. Quality of life issues for survivors and the potential for radiation-induced secondary conditions such as cardiac toxicity and secondary cancers need to be minimized. Dose limits are inappropriate in such circumstances, but detailed dose evaluations are performed in cases of medical exposure. An additional aspect of optimization in medical exposure is the minimization of the probability of an accident—be it underexposure in therapeutic procedures leading to reduction of outcomes such as tumor control probability or overexposure in any procedure leading to the risk of intolerable side effects. It is usually assumed that a 5% discrepancy in radiation therapy dose results in a clinically significant difference in outcome.[4] This tight margin makes optimization essential.

For a person assisting a patient—also a category considered a medical exposure—there may be additional measures employed to minimize dose, including PPE to shield parts of their bodies. In many jurisdictions, medical exposures also cover voluntary participants in medical research and are a matter that is incorporated into local regulations that prescribe dose constraints that differentiate between low-risk activities and higher-risk activities. In these cases the considerations must be a balance of risk and benefit. Clearly two distinct scenarios exist:

1. The volunteer has a (potential) benefit from the research. An example would be additional diagnostic procedures that are not standard practice.
2. No obvious benefit for the volunteer involved in the research. In this case the societal benefit must be weighed against the risk for the individual.

In any case it cannot be for the individual researcher to decide if an exposure is justified and a human ethics committee (also termed institutional review board [IRB]) must be involved. These issues have also been dealt with in ICRP62.[5]

Occupational exposures not only include exposures to irradiating apparatus and artificial radionuclides, but may involve situations where workers are exposed to elevated levels of "natural" radiation, including aircraft and space flight, as well as exposures to naturally occurring radioactive materials (NORM) in higher concentrations than the natural environment. For occupational exposures, a full range of exposure mitigation options is available, but in particular the use of PPE should be limited to exceptional circumstances as routine use of PPE can become extremely onerous, resulting in staff not using the equipment. Similarly, administrative controls should not be onerous as failure of implementation becomes more likely.

While occupational exposure is subject to limitation, the ICRP recommends having an optimization process that would compare doses in comparable situations, with the practice being undertaken as part of a quality improvement cycle. The basic dose limit for occupational exposure is 20 mSv/year in ICRP guidance.

In most jurisdictions, an employer is required to maintain records of doses for the entire duration of a worker's employment, and those records need to be transferred to a subsequent employer or the employee when the employment situation ends. This ensures that dose limits are applied to individuals as per the intention of the recommendations.

5.2.2   Engineering Considerations

From a practical perspective, radiation exposure can be reduced in three very basic ways: time, distance, and shielding. In this context, time refers to the exposure time. Exposure time is directly proportional to the dose received. For radiation workers, this means that exposure should be kept as brief as possible, and is of particular relevance when working with radioactive isotopes. In the context of linear accelerators, there is typically no option to reduce the exposure time for staff. When time cannot practically be reduced without detriment to the practice, the engineered controls of distance and shielding must be put in place.

5.2.2.1 Supervised and Controlled Areas

ICRP Publication 73 on radiation protection and safety in medicine[6] outlines practical methods of protection, including the use of supervised and controlled areas. Controlled areas are areas of high dose rate or where potential exposures are high. In controlled areas, workers are required to follow well-established procedures and practices aimed at controlling exposure. Depending on the magnitude of the hazard, the establishment of procedures may involve the use of simulators or benign mock-ups of the practice to allow the dose to be minimized through proficient use of equipment. Controlled areas are designated by the use of signs, and only authorized staff are admitted to perform documented procedures. In later chapters, the application of controlled areas specific to each practice is described in detail. Dose constraints associated with controlled areas vary with local regulation, but the constraint is typically at levels of around 5 mSv/year.

Supervised areas are those where working conditions are kept under review, but special procedures are not required. Supervised areas are also designated by signs. Workers who operate in supervised areas are normally designated as such and have appropriate monitoring. Supervised areas typically are those where workers might receive >1 mSv/year.

Unsupervised areas should be consistent with dose rates or the risk of contamination in the range acceptable for public exposures.

5.2.2.2 General Principles for Shielding of Radiation Facilities

The most common engineered approach to the reduction of occupational and public dose is through appropriate radiation shielding. The following discussion describes the general principles of radiation shielding; specific considerations for shielding of radiology, nuclear medicine, external beam radiotherapy, and brachytherapy facilities are discussed in later chapters.

There are two main methods of dose reduction in a radiation facility: by distance (assuming a $1/d^2$ reduction of dose) and by placement of a physical attenuating radiation barrier. In the design process, distance as a shield is usually inflexible as facilities are often either placed into existing buildings or planned in new buildings where dose reduction with distance alone is impractical. Typically, the shielding designer will be given an existing architectural plan of the facility, a description of the practice to be undertaken, and a description of the usage of the surrounding facilities. From this information radiation doses can be calculated for the facility's surrounds using distance as the sole dose reduction means, and the designer can then use published data to calculate the amount of attenuating material required on walls, floors, ceilings, and so on, to further reduce the surrounding doses to acceptable levels.

Calculation of the radiation dose in areas surrounding the equipment comprises the separate calculation of the primary dose (radiation directly emitted from the equipment as an essential part of its intended purpose) and calculation of the secondary dose (radiation scattered from the patient plus leakage radiation coming from the equipment, i.e., emanating from the equipment but not part of the primary beam). The amount of radiation produced by the equipment depends mainly upon its workload, W, the calculation of which differs between equipment types, and the reader is directed to the later chapters of this book for specific details. In some equipment such as cobalt-60 units and brachytherapy units, leakage radiation can emanate from the equipment even when not in use. Another factor in calculating the radiation dose in an adjoining location is the use factor, U, which refers to the fraction of time that a machine emits radiation in a certain direction or for a certain procedure (e.g., an x-ray unit in which 20% of the exposures are for chest x-rays will result in a use factor of $U = 0.2$ for the chest bucky wall). Again, the user is directed to later chapters for specific equipment use factors.

The radiation dose received by persons in adjoining locations also depends upon the amount of time in which they occupy the particular location. To account for this, an occupancy factor, T, is introduced. For controlled areas the occupancy factor is always $T = 1$ regardless of the amount of time that an individual is present, but for other areas it is determined by the location's use. For example, an office is likely to be occupied by a single person for a full working week, and hence has an occupancy factor of $T = 1$, whereas a corridor is not normally occupied by any one person for more than a fraction of the working week and is accordingly assigned a much lower occupancy factor. NCRP147[7] and NCRP151[8] provide suggested occupancy factors for various locations for cases when occupancy data are unknown.

5.2.3   Quality Assurance of Equipment and Procedures

Fundamental to providing a radiologically safe environment in a medical facility is appropriate testing of new equipment and an ongoing quality assurance (QA) program for the irradiating equipment once in use. If the equipment operates differently from its intended use, not only will the radiation dose to the patient and quality of the procedure possibly be compromised, but the effectiveness of radiation protection measures such as shielding may also be reduced.

The best opportunity to deal with radiation protection issues of equipment is in the planning and acquisition stage, when specifications can be drawn up and agreed upon. Also, building and room design can be optimized prior to installation. When a new item of equipment is acquired it must undergo acceptance testing and commissioning prior to clinical use. Acceptance testing is performed in collaboration with electronics engineers and usually a representative of the manufacturer to ascertain that the treatment unit:

- Is safe to use
- Performs to its specifications

At the time of acceptance testing any issues can be raised and must be addressed by the manufacturer.

Commissioning is the process of acquiring all the data from the linear accelerator that are required to make it clinically usable in a specific department. Therefore, the commissioning procedure will depend on clinical requirements in a particular center and other equipment within an organization. There are many guidelines available for commissioning of equipment, and these are discussed in more detail in the context of their respective applications in Chapters 6 to 9.

Whatever equipment or procedures are in use the commissioning provides assurance that equipment and procedures meet the expected needs. The commissioning process also provides baseline figures that must be checked on an ongoing basis in a quality assurance program. According to the International Standards Organization (ISO) quality assurance consists of "all those planned and systematic actions necessary to provide confidence that a product or service will satisfy given requirements for quality."[9]

Radiation safety should not only be considered when new equipment is installed, but should also be considered when adding modifications to equipment, extending its use, or introducing new practices. When a new practice is to be adopted it must be justified as providing sufficient benefit to the exposed individuals or to society to offset radiation detriment. An assessment of the sources of exposure and pathways to humans must be performed so that a process of optimization can be undertaken to minimize the collective dose. The results of the assessment, including the need for any new personal protection and monitoring, emergency, or other procedures, need to be incorporated into the radiation safety program.

The International Atomic Energy Agency (IAEA) requires that for medical exposures a QA program be in place, in which qualified experts must participate. The IAEA states that QA programs must include not only measurements of the physical parameters of radiation equipment, but also verification of "appropriate physical and clinical factors used in patient diagnosis or treatment,"[10] that is, the data obtained or used for operation of the machine. This section provides a brief overview of the general principles of a QA program in a medical environment. Chapters 6–9 discuss quality assurance issues directly related to each subject area. A QA program is normally introduced into a jurisdiction under a code of practice, for example, the IAEA Code of Practice TRS398 for standardization of absorbed dose measurements in water from radiotherapy equipment.[11] The following discussion is based on the *Basic Safety Standards* series published by the International Atomic Energy Agency[10] and the ISO9000 series maintained by the International Organization for Standardization.[9]

Typically, a QA program will consist of the following:

- A **QA committee** whose membership represents the many disciplines within the department and ideally chaired by the head of department. At the very least, in a medical facility that uses radiation, the committee should include a medical doctor, a physicist, an operator of the equipment (technologist, radiographer, radiation therapist, etc.), and an engineer responsible for service and maintenance. The membership should be appointed and supported by senior management, and members must have sufficient depth of experience to understand the implications, and have the authority to instigate and carry out the QA process. The committee should be visible and accessible to staff and is responsible for initiating and tailoring a program to meet the needs of the department as well as monitoring and auditing the program once it is in place. The committee must also have a terms of reference, meet at established intervals, and retain minutes of its meetings for audit purposes.

- A **policy and procedures manual** that contains clear and concise statements about responsibilities and all practices undertaken within the department. The manual should be reviewed regularly and updated as procedures change. The manual must be approved by the head of department and appropriate section heads, but it is also important that all staff have input into it and agree on its contents. A record should be kept of the location of all copies of the manual to ensure that each copy is updated when required. As a minimum, the manual should include sections on administrative procedures, clinical procedures, treatment procedures, physics procedures, and radiation safety. In the context of providing documentation to staff it is essential to consider the use of appropriate language, and it is often required to have translations of important radiation protection documents in commonly spoken languages. This also applies to information for visitors and patients (e.g., about pregnancy and radiation).

- A **quality assurance team** with a well-defined responsibility and reporting structure and consisting of members from all disciplines within the department. Each member of the team must know his or her own responsibility, be trained appropriately, know what actions are to be taken (and understand the consequences) should a test or action be outside the preset action levels, and maintain records documenting any corrective action taken.

- **Appropriately qualified staff and the provision of training** are an essential part of the provision of quality services. In the case of radiation protection, training should be provided to all staff, and most jurisdictions require documentation that persons who are classified as occupationally exposed undergo regular radiation protection training.

- A system of **quality audits** that involves a review of the compliance of activities with planned arrangements and whether the arrangements are implemented suitably to meet the objectives. The audit should ideally be performed by somebody from outside the organization, for example, the IAEA/WHO thermoluminescent dosimeter program for checking of dose from radiotherapy units.

Paramount to all the above is documentation of all activities. One key component to an ongoing QA program is action levels, which are set by the QA committee. An action level is a quantitative point at which an intervention is required. For example, in a radiotherapy department the physics section is given the authority to ensure that the radiation output of the linear accelerators is correct. A two-phase action level in this instance might include:

- "For any daily radiation measurement that exceeds 2% but less than 4% of expected, treatment may continue but the senior physicist responsible must be notified immediately."

- "For any daily radiation measurement that exceeds 4% treatment must stop immediately and the problem be investigated by the senior physicist."

For the QA system to be successful action levels must:

- Be quantitative
- Reflect the required outcome
- Be informed by the achievable outcome
- Be unambiguous
- Be easy to understand

The QA committee should conduct a review when an action level has been exceeded or where set procedures have been discovered to be faulty. After the review recommendations must be formulated in writing for improvement of the QA system.

Finally, quality assurance includes a system of reporting. This applies to reports both within an organization and outside. In the context of radiation protection, reporting responsibilities are typically clearly specified in the conditions for licensing.

5.3 Monitoring

There are many situations in which monitoring is required to confirm that radiological measures are adequate or that controls have been complied with. Monitoring situations include:

- General external gamma-ray exposure
- Extremity exposures
- Contamination monitoring
- Leakage from irradiating apparatus
- Neutron monitoring

Whatever monitoring is used must be fit for the purpose. All too often, a manager will insist on the use of a film badge or thermoluminescent dosimeter (TLD) for all staff in an area, just to ensure that everything is okay without any appreciation of the true hazard or the limitations of such monitors. Of course, in a situation involving radioactive material or a neutron field, such a monitor may be totally insensitive to the hazard. Examples of some monitoring devices are shown in Figure 5.1.

5.3.1 External Gamma-Ray Exposure

In many cases, external gamma-ray exposure is the primary radiation hazard; it is typically monitored through the use of personal dosimeters, for example, film badges or thermoluminescent dosimeters, optically stimulated devices, and pocket electronic dosimeters. There is a range of monitoring services available using these dosimeters, with LiF- or $CaSO_4$-based TLDs being the most common. Monitors should be chosen to match the dose range expected, which usually mitigates against the use of LiF, which is less sensitive than $CaSO_4$, although the use of $CaSO_4$ monitors may cause some difficulties if the radiation spectrum is unknown, as it is far from tissue equivalent. These monitors are designed to give an estimate of whole body exposure and are normally worn at chest or waist level.

FIGURE 5.1
Examples of personal monitors: quartz fiber, TLD, and film.

Where the worker uses PPE, the dose measurement should reflect the protection afforded by the PPE. For example, if an employee is required to wear a lead gown during the course of his or her duties, the dosimeter should be worn under the gown.

Personal monitors for gamma-ray exposure are normally issued with a control dosimeter to measure background radiation received by the wearer. This control dosimeter must be stored in an area where only background radiation will be measured.

For fixed installations, dose rate surveys or radiation leakage tests may be more effective for personal protection than individual monitoring.

5.3.2 Extremity Dose

Doses to extremities are often monitored in situations where small high-activity sources are handled such that dose to the extremity (usually fingers) greatly exceeds the whole body dose. These are usually TLD powders contained within a ring (see Figure 5.2) or in sachets that can be taped to fingers.

5.3.3 Radioactive Contamination

Radioactive contamination is typically monitored by surface contamination monitoring and air sampling where the hazard is airborne. In the case of contamination, radioactivity might be directly detectable through the use

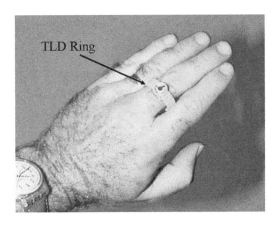

FIGURE 5.2
Example of a TLD ring dosimeter.

of alpha- or beta-ray contamination monitors; however, in some cases swipe sampling and radionuclide assay may be needed. These techniques are necessarily indirect, and there are many input parameters needed to model the dose. These parameters include the chemical form of the radionuclide, particle size for the airborne particle, and equilibrium within the decay chain involved. If the potential risk of contamination is sufficiently high, it may be necessary to involve in vivo measurement techniques for workers, for example, whole body monitoring, thyroid measurements, or the sampling and analysis of excreta.

5.3.4 Environmental Monitoring

In environmental situations, monitoring can be very complex and may require knowledge of diet, chemical forms of radioactivity, transfer factors from food and the environment to uptake to humans, dust loadings, particle size data, attached/unattached fraction for Rn daughters, equilibria of decay chains, and external beta- and gamma-ray dose rates. All of this may need to be differentiated from the preexisting situation before the practice commenced, as it is only the increase in dose from the practice that is subject to the system of radiation protection.

5.3.5 Neutron Monitoring

In the use of higher-energy electron linear accelerators among other potential hazards in medicine (e.g., positron emission tomography [PET] cyclotrons), neutrons can be a hazard. Neutrons tend to only become a hazard in situations involving fixed installations. Normally neutron fields would be measured with survey monitors, although ^{6}LiF TLD detectors are available for personal monitoring as well as bubble dosimeters. Personal monitors for neutrons tend to have low sensitivity.

References

1. ICRP, *1990 Recommendations of the International Commission on Radiological Protection*, ICRP Publication 60, Oxford: Pergamon Press, 1990.
2. ICRP, ICRP Publication Number 103: Recommendations of the ICRP. 2007: Elsevier.
3. Clarke, R., 21st century challenges in radiation protection and shielding: Draft 2005 recommendations of ICRP, *Radiation Protection Dosimetry*, 2005, 115: 10–15.
4. ICRU, *Determination of Absorbed Dose in a Patient Irradiated by Beams of X or Gamma Rays in Radiotherapy Procedures*, ICRU Report 24, Washington, DC: ICRU, 1976.
5. ICRP, *Radiological Protection in Biomedical Research*, ICRP Publication 62, Oxford: Pergamon Press, 1993.
6. ICRP, *Radiological Protection and Safety in Medicine*, ICRP Publication 73, Amsterdam: Elsevier, 1996.
7. NCRP, *Structural Shielding Design for Medical X-ray Imaging Facilities*, NCRP Report 147, Bethesda: MD: National Council on Radiation Protection and Measurements, 2004.
8. NCRP, *Structural Shielding Design and Evaluation for Megavoltage X and Gamma Ray Radiotherapy Facilities*, NCRP Report 151, Bethesda, MD: National Council on Radiation Protection and Measurements, 2005.
9. ISO, *Quality Management and Quality Assurance Standards*, ISO9000 Series, Geneva, Switzerland: International Organization for Standardization.
10. IAEA, *International Basic Safety Standards for Protection against Ionizing Radiation and for the Safety of Radiation Sources*, Safety Series, Vienna, Austria: International Atomic Energy Agency, 1996.
11. IAEA, *Absorbed Dose Determination in External Beam Radiotherapy: An International Code of Practice for Dosimetry Based on Standards of Absorbed Dose to Water*, Vienna, Austria: International Atomic Energy Agency, 2000.

6

Radiation Protection in Diagnostic and Interventional Radiology

John Heggie

CONTENTS

This chapter addresses those protection issues that arise specifically in the context of diagnostic and interventional radiology. The protection of both the patient and occupationally exposed individuals is addressed in some detail, and the protection of members of the public in general is considered in the context of site shielding. In order to fully appreciate how protection may best be optimized, a few introductory comments will be made about the production of x-rays.

6.1 Introductory Comments about the Type of Radiation Involved, Its Production, and Its Use

6.1.1 Production of Diagnostic X-Rays

The primary mechanism for the production of x-rays is as a result of the deceleration (or acceleration) of a rapidly moving charged particle as noted in Chapter 2. Radiation produced in this manner is known as *bremsstrahlung*, after the German term for braking radiation. When a charged particle, such as an electron, passes through matter, it interacts with atomic nuclei via the Coulomb force. The electron is deflected from its original direction and loses energy that is emitted directly as electromagnetic radiation. Generally the charged particle penetrates many atomic layers giving up only a very small fraction of its total energy in each of several interactions before it ultimately comes to rest. Occasionally, a head-on collision with the nucleus will result in the production of a photon whose energy is the same as that of the incident charged particle energy.

In the usual situation the charged particle will undergo numerous interactions before coming to rest, and the energy given up in each collision is sufficiently

FIGURE 6.1
A typical dual-focus, rotating-anode x-ray tube for use in diagnostic radiology. The key features are identified.

small that very few x-rays are in fact produced. At radiation energies typical for radiology, most of the energy (>99%) is dissipated as heat as the charged particle produces numerous ion pairs before coming to rest. Those x-rays that are produced will have a continuous spectrum of energies ranging from essentially zero to the maximum energy that the charged particle carries.

In most x-ray tubes used in diagnostic radiology, electrons are accelerated toward a tungsten anode (target) (see Figure 6.1) by applying a large accelerating voltage between the anode and the cathode. In all but the most recently designed generators the accelerating voltage is not kept constant with time but will vary up to a peak voltage designated kVp. In so-called single-phase x-ray generators the voltage swing may be as high as 100%, ranging from zero to the kVp. After acceleration at impact with the target, the electron will have an amount of energy that is proportional to the instantaneous applied voltage; thus, very few electrons acquire a kinetic energy numerically equivalent to the kVp applied to the tube. As noted above, even fewer x-rays are emitted with this energy since the bremsstrahlung process generally involves the production of a large number of low-energy photons rather than the emission of a single photon with energy equal to the incident electron. Thus, the bremsstrahlung spectrum will be continuous with all energies present up to a maximum energy determined by the maximum accelerating voltage applied to the tube (see Figure 6.2). In summary, the continuum results because of:

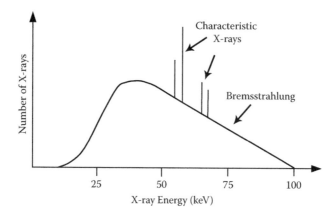

FIGURE 6.2

Simplistic representation of the spectrum obtained using a diagnostic x-ray tube operated at 100 kVp. The characteristic x-rays from the tungsten anode target are superimposed on the continuous bremsstrahlung radiation spectrum. The low-energy x-rays have been preferentially removed by filtration.

- The different energies with which the electrons reach the target
- The variation of the Coulomb force with the distance of the electron from the nucleus center
- The fact that the electrons undergo multiple collisions dissipating differing amounts of energy in each process

There is a second process by which x-rays can be produced, and that occurs when the incident electron has sufficient energy to eject an electron from an inner shell of the target atom. Removal of such an electron leaves the atom with net positive charge and an electron vacancy in an inner shell. The atom will attempt to restore itself to its lowest possible energy state, and this is achieved by a higher shell electron dropping into the vacant orbital with the *simultaneous* emission of a photon of radiation. The emitted photon, which may be sufficiently energetic to be classified as an x-ray, has an energy corresponding exactly to the difference in energy between the two shells and is unique to the particular target material—hence the name *characteristic radiation*. The new vacancy created in the higher shell will be filled by a second electron originating from an even higher shell with the emission of a second characteristic x-ray. This x-ray will have substantially less energy than the first. The whole process is repeated until the atom is totally de-excited. In the case of the tungsten anode x-ray tube, provided the kVp exceeds the K-shell binding energy of 70 keV, there will be up to four different K-shell characteristic x-rays superimposed on this continuum. The contribution of characteristic radiation relative to the bremsstrahlung depends on kVp and filtration in the useful beam and, with the exception of mammography tubes, is not usually dominant, being typically between 10 and 30% of the total intensity.

6.1.2   Energy Range of Relevance

Typically general-purpose x-ray tubes for diagnosis are designed to operate at tube potentials anywhere between approximately 40 and 150 kVp, with most applications performed at 100 kVp or less. There are a few exceptions where atypical tube potentials may be used. For example, the unique demands of mammography require that x-ray tubes are frequently operated between potentials of 25 and 30 kVp and are rarely operated above 35 kVp, although the recent move to digital mammography technology has meant that slightly higher tube potentials may be utilized satisfactorily. On the other hand, computed tomography (CT) technology usually requires higher tube potentials, and the minimum tube potential encountered is 80 kVp, with most CT tubes operated at just a few discrete values, such as 80, 100, 120, and 140 kVp. The actual spectrum shape depends on the type of generator (single phase, three phase, constant potential, etc.) and, most significantly, on the amount of filtration that may be placed in the beam. This has particular significance in the context of optimizing radiological procedures, as will be discussed later (see Section 6.2.1.2).

6.1.3   X-Ray Tube Shielding and Primary Beam Definition

The x-rays produced by any x-ray tube are radiated in all directions so that in the absence of any tube shielding a significant hazard exists for staff and patients. Fortunately, the modern x-ray tube is invariably shielded with lead to reduce the intensity of unwanted radiation in all directions save for a small collimated area, the *primary beam*, that is used to irradiate only the part of the patient that is of clinical interest. The boundary of the primary beam may be defined by fixed cones, as in some mammography units; by adjustable slits, as in CT equipment; or by continuously variable multileaf lead collimators, as in general and fluoroscopy equipment. The collimator frequently incorporates a light source and mirror system so the adjustment of the x-ray beam to the required field of view (FOV) is achieved by noting the margins of a light field on the patient (see Figure 6.3). However, even with x-ray tube shielding and lead collimators, there remains some *leakage radiation* that penetrates the lead housing and collimators. In most applications and in most, if not all, jurisdictions, the leakage radiation, expressed as the air kerma rate, must be limited by design to a value of less than 1 mGy per hour averaged over an area of 100 cm², with the x-ray tube operated at its maximum tube potential and maximum continuous tube current. For dental x-ray tubes the allowed leakage radiation air kerma rate is less than 0.25 mGy per hour.[1]

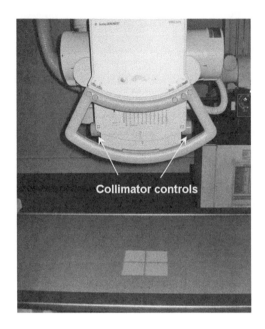

FIGURE 6.3
A typical multileaf collimator showing the light beam collimator controls (arrowed).

6.2   Protection of the Patient

6.2.1   General Protection Principles

6.2.1.1   *Justification*

All potential diagnostic and interventional radiology exposures must be subject to the principles of justification and optimization. The justification principle is common to all practices dealing with exposures to ionizing radiation and may be stated as follows:[2]

> All exposures must show a sufficient net benefit when balanced against any possible detriment that the dose might cause.

For patients undergoing medical diagnosis or treatment, there are two levels of justification. First, the practice involving exposure to radiation must be justified in principle. For example, the well-established practice of taking daily chest radiographs of patients who have recently undergone cardiac bypass surgery is generally acknowledged as warranted. However, there are newly evolving screening practices for which a net benefit has not been demonstrated at this time, for example, CT cardiac scoring, CT screening of high-risk groups for lung disease, and whole body three-dimensional CT scanning. Such practices are not currently endorsed by the relevant professional medical bodies, and in this context, the continuing involvement of medical professional societies should be ensured, as matters of effective medical practice will be central to this judgment.[3] Second, each procedure should be subject to a further, case-by-case justification by both the referrer, who is responsible for the management of the individual patient and who determines that the exposure is necessary for diagnostic purposes, and the radiologist or other practitioner who may direct the radiological procedure.[4]

Ultimately, the decision to perform a radiographic or interventional examination rests upon a professional judgment of the benefits that accrue to the total health of the patient, as opposed to any biological effects that might be caused by the ionizing radiation. The benefit will be the potential diagnostic information or therapeutic effect of an interventional procedure resulting from the medical exposure, including the direct health benefits to an individual as well as the benefits to society. The detriment will be the potential deleterious effects of ionizing radiation. In the doses generally arising in diagnostic imaging the only possible detriment is the small probability of inducing a cancer or a genetic mutation, which may be passed on to future offspring (compare also Chapter 3). These effects are generally thought to be stochastic in nature, meaning that the probability of occurrence increases with increasing dose and there is no safe threshold below which they cannot occur. The age of the patient and the anatomical region examined are also relevant in the risk assessment. For interventional radiology procedures, an additional concern relates to possible deterministic effects such as skin damage.[5] In these instances, the severity of the effect increases with increasing dose, and there is usually a threshold below which they do not occur.

The justification process should also take into account the efficacy, benefits, and risks of using alternative imaging modalities involving no or less exposure to ionizing radiation, for example, ultrasound, magnetic resonance imaging, and endoscopy.[3,4] Also influencing this choice will be practitioner preference and expertise and the availability of the differing imaging modalities.

6.2.1.2 *Optimization*

Once clinically justified, each examination should be conducted so that the dose to the patient is the lowest necessary to achieve the clinical aim. The quality of the images and the complexity of the examination should be sufficient for the intended purpose of the procedure. Since patients may accrue direct benefits from medical exposures, it is not appropriate to impose strict limits on the doses received from fully justified examinations. However, patient dose surveys indicate wide variations in delivered dose to achieve satisfactory image quality, indicating that there is significant scope for the implementation and optimization of patient protection.[6] To this end, it is recommended that diagnostic reference levels (DRLs) be established as a practical tool to aid in dose optimization (see Section 6.2.1.4).

In general the optimization process necessarily requires a balance between patient dose and image quality. Dose reductions must not be achieved without regard to any loss of diagnostic quality in the image that may accompany the dose reduction. Images of unacceptable quality may result from unwarranted reductions in patient dose rendering the images undiagnostic and ultimately leading to repeat examinations and higher patient doses. The requirement for image quality should be tailored to the clinical problem, and lower levels may be acceptable in some circumstances. Certainly, the size and shape of the patient will influence the level of dose required. Accordingly, the operator of radiographic equipment must minimize patient dose under the constraint that the image quality is acceptable for the diagnostic information sought.

A crucial cog in the optimization process is the provision of appropriate training for operators of x-ray equipment so that they may fully understand the potential of the equipment. It would be anticipated that radiation health professionals (radiologists and radiographers) should be deemed to have significant knowledge by virtue of undertaking a course leading to their professional qualification. Other professional groups (e.g., cardiologists and vascular surgeons) who perform or direct exposures using x-rays should also have appropriate training. This training should include, as a minimum, a knowledge of:

- The key features of the relevant x-ray and ancillary equipment
- Risk factors such as age and tissue radiosensitivity
- Measurement of radiation dose
- The magnitude of typical doses from relevant examinations
- Methods of reducing patient doses during radiological examinations
- Methods for minimizing the occupational hazards arising from the use of x-ray equipment
- Occupational dose limits and the ALARA principle

Regardless of professional qualifications, additional training specific to the equipment used at a particular institution should be provided by the practice. In some instances, most notably with CT and interventional equipment, training at commissioning may be provided by the equipment supplier's representative.

The use of mobile radiographic equipment should be kept to the minimum. Procedures should be performed with *fixed* radiological equipment when possible, since they offer a wider choice of technique factors and superior ability to establish the correct geometrical relationship between x-ray tube, patient, and imaging device, when compared with mobile units. In addition, they invariably offer greater protection for the operator and other patients. If the use of mobile fluoroscopy equipment cannot be avoided, as in the operating theatre environment, it is recommended that they be equipped with automatic brightness control (ABC) and image storage facilities so that last image hold (LIH) and other image processing techniques may be employed.

With these general principles in mind there are a number of precautions that the operator of any x-ray equipment may take in order to optimize any diagnostic or interventional radiological procedure. For example, the operator should:

- Tailor the technical factors for the radiological procedure to the patient's specific anatomy. For example, it is recommended that the highest kVp be used that is compatible with adequate image quality (contrast). This does increase the scatter as a fraction of the patient skin dose, but the latter is reduced dramatically. Likewise, increasing the beam filtration will reduce the patient skin dose by preferentially removing low-energy x-rays from the primary beam. These low-energy x-rays contribute significantly to patient dose and only minimally to the image formation, so that using extra filtration, within limits, has minimal impact on image quality. For example, most bodies recommend that the minimum total aluminum equivalent filtration in the beam for general radiography should be 2.5 mm. However, even radiography of extremities may be undertaken with at least 4 mm of aluminum total filtration equivalent in the beam without unduly compromising image quality.

- Restrict the number of views (or images) per examination to the minimum necessary. For example, question whether the lateral view in chest radiography is always required. Are plain skull films necessary as an adjunct to a CT scan when the latter will almost certainly provide superior diagnostic information? Are both contrast and noncontrast CT examinations required?

- Choose the most efficient image receptor required to achieve the diagnostic information. This may be as simple as choosing a fast versus a slow intensifying screen provided the image noise is acceptable, and ensuring that the film and screen combinations are appropriately matched. Other than for radiography of extremities and mammography, and perhaps pediatrics, there are few occasions when film-screen systems with a nominal speed of less than 400 are warranted.

- Avoid the universal use of high-ratio antiscatter grids or grids at all, most particularly in the context of radiography and fluoroscopy of children.

- Ensure that the primary x-ray beam is collimated to within the size of the image receptor in use and then only to the clinically relevant FOV. This has the added benefit of simultaneously improving image quality by reducing impact of scatter radiation as the amount of scatter increases markedly with increasing field size.

- Use carbon-fronted cassettes, carbon fiber tabletops, carbon- or aluminum-fronted image intensifiers, and flat-panel imagers to minimize absorption of radiation between the patient and the image receptor. The effectiveness of carbon-fronted cassettes has been investigated by Dance et al.,[7] and their use, when compared with aluminum-fronted cassettes, has been shown to offer clear dose and contrast advantages across a broad range of techniques.

- Avoid the use of extremely short source (focus)-to-image distances (SIDs), as this can lead to unnecessarily high skin doses and is particularly pertinent in fluoroscopy and interventional procedures with C-arm or U-arm equipment. Most x-ray equipment, with the exception of mobile x-ray units, is designed to prevent the SID from being less than 200 mm.

- Shield radiosensitive organs such as the gonads, lens of the eye, breast, and thyroid whenever feasible, particularly when children are being irradiated. However, it should be appreciated that protective drapes do not guard against radiation scattered internally within the body and only provide significant protection in cases where part of the primary x-ray beam is directed toward structures outside the immediate area of interest. In general, shielding should be placed on the patient surface facing the primary beam. In CT examinations, given the cylindrical symmetry of the exposure, to be effective the shielding must be wrapped around the patient. Gonadal shielding is most useful in males if the gonads are in the primary beam, but is less useful in females because of the substantial amount of internal scattering that contributes to the absorbed dose received by the ovaries. Often, it is not possible to shield the ovaries because any shielding may interfere with the diagnostic information sought, and in any event there is enormous variation in their anatomical location.

- Where relevant ensure that the film processor function (e.g., chemistry, developer temperature, replenishment rate, and dwell time) is optimized according to the manufacturer's recommendations. Surveys of radiological practices have found a significant spread in processing speed.[8,9] This is not surprising since system speed, contrast, and fog are very dependent on temperature and dwell time. Routine sensitometry/densitometry conducted as part of a comprehensive quality assurance (QA) program (see Section 6.2.1.3) will assist in establishing optimal processor performance.

- Avoid repeat procedures. If a radiographic procedure needs to be repeated, this of necessity will result in unnecessary exposure to both the patient and the operator. Repeat exposures may be necessary due to the poor quality of the image or if the image does not provide the clinical information required. The latter cause can be avoided by the careful planning of the examination to fit the clinical problem. Care is also necessary to ensure the correct positioning of the patient with respect to the image receptor and x-ray tube. Repeat exposures due to technical errors can be minimized by the correct selection of exposure factors consistent with the region being examined, the speed of the image receptor, and processing procedures when relevant. If automatic exposure control (AEC) is not available, it is recommended that technique charts be posted for the common radiographic examinations to assist in maintaining proper image quality. A comprehensive QA program, which includes reject analysis, should highlight systematic errors or problems and ultimately lead to a lower repeat rate. In any event, repeat exposures should not be undertaken simply because an image may not be of the highest quality. If the image contains the required information, then a repeat should not be performed.

6.2.1.3 Quality Assurance

A key element in radiation protection of patients (and occupationally exposed staff) is the establishment of a quality assurance (QA) program with particular emphasis on image quality optimization and patient dose reduction as its primary and secondary goals, respectively. The basic elements of the QA program might include acceptance testing, constancy testing, record keeping, and patient dose surveys (see Section 6.2.1.4).

At initial installation, the diagnostic or interventional radiology equipment and its associated equipment (e.g., film processors, computed radiography equipment) should undergo a series of acceptance tests. Some suggestions as to the type of testing that may be undertaken can be obtained by reference to the relevant national and international standards (e.g., reference 10) and to publications by professional bodies (e.g., references 11–13). These tests may be used to verify, or otherwise, that the initial performance of the equipment conforms to the manufacturer's specifications and to the standards. The results of the acceptance tests should be used in part to define the acceptable range of parameters that will be monitored in any subsequent constancy testing and should be thoroughly documented.

Following acceptance, constancy tests designed to assess the subsequent performance of the equipment should be performed. These are usually less involved tests that may be performed by radiographers and are designed to assess image quality and patient dose. As such it is recommended that system tests using appropriate image quality phantoms form the basis of constancy tests. It should be noted that the International Commission on Radiological Protection (ICRP) has identified[4] the need for different constancy and acceptance tests to be performed when digital radiography equipment is utilized. The QA program should outline the types of constancy tests, frequency of tests, the tolerance of each parameter monitored, and the procedure for staff to follow when tolerances are exceeded. The results of constancy testing must be recorded and reviewed as a matter of routine, and any anomalous results reported immediately to the person responsible for the QA program management. In particular, failures identified during acceptance or constancy testing, and actions taken to remedy these failures, should be thoroughly documented.

In extreme instances, when the results of constancy tests indicate that the equipment is outside tolerance, the results may be used to justify replacement of equipment. For example, the efficiency of an image intensifier deteriorates with time because of loss of vacuum and radiation damage to the output phosphor. At some point this loss of efficiency will be sufficiently severe that the dose to the patient will exceed acceptable levels and replacement is warranted.

An analysis of the reasons for rejecting images, whether produced on film or in digital form,[14] is a fundamental aspect of the QA program and should be undertaken by a senior radiographer. Errors of positioning and image labeling may emerge that can be remedied by appropriate instruction. Over- or under-exposure errors may be indicative of a fault with a particular x-ray tube in a particular room or point to a case of mismatched film-screen combinations, for example. It is important to note that reject analysis should be conducted as part of an educative rather than a punitive process. Cooperation, not alienation, of radiographers and others is a key to a successful QA program.

6.2.1.4 Diagnostic Reference Levels

Dose limits are not appropriate for patients undergoing diagnostic or interventional radiology procedures.[15] However, as part of good radiological practice, patient dose surveys should be undertaken periodically to establish that doses are acceptable when compared with recommended values of diagnostic reference levels (DRLs). A DRL is a reference level of dose (e.g., entrance surface dose) likely to be appropriate for average-sized patients undergoing medical diagnosis and treatment.

It is suggested that institutions establish their own local DRLs if national values are not established, and patient doses should be compared with these values at appropriate intervals.[16] Ultimately, DRLs should be established for both adults and pediatric patients at the national level for most common examinations, by the relevant professional societies in consultation with regulatory authorities. Any local DRL should be set with due regard to these national DRLs where they are available. For adults, the DRLs are usually defined for a person of average size, which is taken to be about 70 to 80 kg, and the recommended values are frequently chosen as a percentile point (typically the 75% level) in the observed distribution of doses to such patients (see Figure 6.4). Accordingly, when performing dose surveys, patients within this weight range should be selected. It should be appreciated that DRLs do not represent best practice, and the ultimate target for any institution should be to lower their doses to a level regarded as achievable. For any procedure, an achievable dose is one that maximizes the difference between the benefit and risk without compromising the clinical purpose of the examination.[6] DRLs are also not set in stone and should be reviewed and adjusted, by the relevant regulatory authorities in consultation with the relevant professional

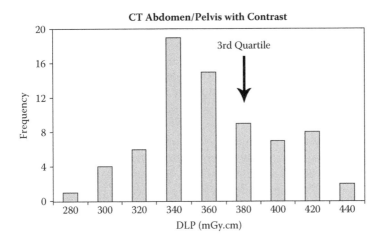

FIGURE 6.4
Establishing a local DRL for a CT examination. The histogram plot is for a sample of seventy-one adult patients who underwent routine CT abdominal/pelvic examinations with contrast. The third quartile value of the DLP is approximately 380 mGy·cm, which may be rounded up to 400 mGy·cm for use as a local DRL.

societies, at intervals that represent a compromise between the necessity for stability and long-term changes in the dose distributions arising from technological improvements. Usually the adjustment produces a lowering of the DRL as a result of technological improvements.

The choice of a dose descriptor to use as a DRL depends on the type of examination. The DRL should be expressed as a readily measurable patient-related quantity for the specified procedure. Usually, for general radiographic examinations it is taken to be either the entrance surface dose (ESD) or the dose area product (DAP), for fluoroscopic examinations it is taken to be the DAP, and for CT examinations it is taken to be the dose length product (DLP).

Practices should review and justify, on the grounds of clinical needs, any dose values that fall significantly above or below the established DRL.[15] DRLs being repeatedly and substantially exceeded may indicate an underlying fundamental problem that warrants investigation. However, DRLs should be applied with flexibility to allow higher doses if these are indicated by sound clinical judgment.[17] Further, as emphasized in Section 6.2.1.2, patient dose surveys must always be undertaken in parallel with image quality assessments. Paying regard to DRLs for common procedures forms a substantive part of the optimization process.

6.2.2 Specific Protection Issues Relating to General and Dental Radiography

Most of the protection issues of relevance to general and dental radiography have been addressed in Section 6.2.1.2 with two possible exceptions. First, the use of physical compression of tissue in some instances (e.g., mammography, intravenous pyelograms [IVPs]) is recommended and represents a rare win-win situation, as this will both improve image quality, by reducing the contribution of scatter to the image formation, and reduce the patient dose as a bonus. Second, automatic exposure control (AEC) technology should be utilized in general radiography whenever possible as its use will aid in reducing the retake rate arising from under- or overexposed images. Further procedure specific advice is available in the European guidelines,[18,19] which have been developed to provide specific advice on good techniques when radiographing children and adult patients, respectively.

6.2.3 Specific Protection Issues Relating to Computed Radiography and Digital Radiography

While the provisions outlined in Sections 6.2.1.2 and 6.2.2 apply equally when computed radiography (CR) and digital radiography (DR) technology are being utilized, there are aspects of their use that are unique. Choosing the appropriate image processing parameters is just one aspect that must be considered. More importantly, from a radiation protection perspective, the user must recognize that these technologies offer a wide dynamic, and there is documented evidence that radiographic staff may produce high-quality images by increasing the radiation dose with the implementation of this technology.[20,21] If conventional film-screen systems had been used, these same images would have been grossly overexposed, but the wide dynamic range of the digital modality allows much more latitude in the choice of exposure factors. This undesirable *exposure creep* may be prevented by utilizing an appropriately adjusted and optimized AEC.[4,22]

6.2.4 Specific Protection Issues Relating to Fluoroscopy

For examinations including barium studies, angiography, and interventional radiology, all of which use fluoroscopy as well as static imaging, the prescription of what constitutes optimal techniques may be difficult to define. In many instances, the conduct of the examination is unique to the patient. However, a list of commandments for reducing patient dose in fluoroscopic examinations has been formulated[5,23–25] and is recommended for adoption. The basic tenets are discussed below.

- Even when using equipment featuring automatic collimation adjustment to the selected field size, some manual adjustment to the FOV is usually warranted.[26] Dose reductions will also arise if the collima-

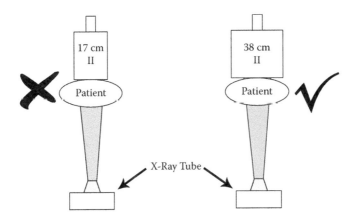

FIGURE 6.5
The impact of using a 17 cm diameter field size versus a 38 cm diameter image-intensifier selected field size, that is, using electronic zoom, may result in a significant dose penalty to the patient. Through appropriate use of collimation the FOV is the same in both instances.

tors are adjusted electronically in the absence of x-rays. More significantly, use the largest field size collimated down to the FOV that is consistent with the imaging needs (Figure 6.5). Restated, that means avoiding the use of electronic magnification. Electronic magnification results in dose rates to the patient that may be several times higher than those that apply when the largest field size is chosen. For example, to achieve comparable image noise and brightness, a nominal 38 cm diameter image intensifier may operate at a dose rate that is approximately one-fifth of the dose rate pertaining to the same image intensifier with a 17 cm diameter field size selected. In practice, with newer image-intensification systems, sophisticated adaptive filtration noise reduction software is usually employed to reduce the noise with the small field size, and the dose rate increase is more modest. Nevertheless, the dose increase with small field sizes remains significant. This also applies to flat-panel technology used for fluoroscopy. An additional benefit accrues to the operator, as his or her exposure will be directly related to the radiation exposure of the patient. Unnecessarily high patient exposure means unnecessarily high operator exposure.

• Always optimize the geometry (i.e., avoid geometric magnification). That is, the patient should be placed as close to the image receptor as possible, and the latter should be moved to the maximum distance from the x-ray source, as illustrated in Figure 6.6. Restated, move the patient to the image receptor and *not* the image receptor to the patient. Reduced patient/staff exposure and improved image quality will result. Good geometry will also improve image resolution by

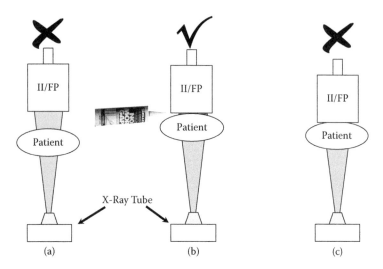

FIGURE 6.6
Impact of poor geometry on patient skin dose during fluoroscopy. In (*a*) there is an undesirable air gap, which has been avoided in (*b*). A skin dose decrease of approximately 30% is indicative only but is typical of operation at a SID of 1 m. In (*c*) the SID has been decreased unnecessarily, resulting in a skin dose penalty of approximately 10% compared with the geometry shown in (*b*).

minimizing the undesirable effects of the focal spot penumbra. In an extension of these concepts, it is advisable to avoid steeply oblique views because a modest 3 cm increase in projected body thickness through using oblique views results in an approximately 100% increase in skin dose and inferior image quality.

- Always utilize automatic brightness control (ABC) and pulsed fluoroscopy, in which the x-ray tube is switched on and off in a regular sequence typically at 12.5 or 15 frames per second, whenever possible. Most manufacturers offer this latter feature as an option now, and typically dose reductions of up to 50% may be achieved. In any event, whenever a choice in dose rate is available, the lowest dose rate commensurate with obtaining adequate image quality should be used.

- Minimize the amount of fluoroscopy, that is, keep the foot off the pedal, taking full advantage of any last image hold (LIH) technology. However, operators should be aware that elapsed fluoroscopy time is not a reliable indicator of dose. Patient size and procedural aspects such as locations of the beam, the use of oblique views, image receptor dose rate, and the number of acquisitions can cause the maximum skin dose to vary by a factor of at least ten for the same total fluoroscopy time. The operator should appreciate that dose rates will be greater and dose will accumulate faster in larger patients. Note, however, that Marshall et al.[27] observed that in complex procedures patient dose is more likely to be affected by operator choices and clinical complexity than by the physical size of the patient.

- Never use fluoroscopy as a convenient substitute for plain radiography. On a per image basis, fluoroscopy may seem to be an efficient use of radiation. However, when due account is taken of the frame rate and the duration of fluoroscopic procedures, it can only be regarded as a high-dose procedure. To put this in perspective, the entrance surface dose (ESD) for an abdominal radiograph is approximately 3 mGy,[22] while the dose rate at the surface of the skin during fluoroscopy with an efficient imaging chain is typically about 30 mGy/min. That is, a mere 6 s of fluoroscopy gives rise to the same ESD as a single abdominal radiograph. In an attempt to limit fluoroscopic doses, authorities regulate to ensure that the maximum air kerma rate, without backscatter, at the position of the skin entrance can never exceed 100 mGy/min under normal conditions of operation (e.g., references 28 and 29).* In other jurisdictions this limit may be as low as 50 mGy/min.[30] Technically, the air kerma rate limitation is achieved by restricting the maximum kVp and mA provided by the x-ray generator or changing the filtration in the x-ray tube housing.

- Most fluoroscopic procedures, such as barium meals and digital subtraction angiography, involve a digital acquisition phase during which static images will be recorded. During this acquisition phase it is important to choose the lowest frame rate and shortest runtime consistent with diagnostic requirements.

- In Section 6.2.1.2 the universal use of grids as a practice was challenged. In the context of fluoroscopy it has been observed by Lloyd et al.[31] that substantial dose reductions of approximately 50% may be achieved in barium enema examinations by not using a grid. These authors have recommended that a grid only be used with obese patients for this examination. In the context of fluoroscopy with children, a grid may be avoided for all but the largest children.[32] Certainly, notwithstanding the previous discussion on optimum geometry (see Figure 6.6), if the image receptor cannot be placed close to the patient, then the air gap that ensues would almost certainly negate the need for a grid.

6.2.5 Specific Protection Issues Relating to Interventional Radiology

There is now a well-documented history of skin damage arising from interventional procedures.[5,23,33–36] The injuries span the whole spectrum from temporary erythema and hair epilation to tissue necrosis, the latter requiring extensive skin grafts over several years. The increasing number and

---

* The FDA requirement given in this reference is actually 10 R/min ≈ 88 mGy/min.

complexity of interventional procedures will, no doubt, exacerbate these problems in the future. Although the majority of interventional procedures are generally for treatment of life-threatening conditions, it is an unfortunate fact that most of these radiation-induced injuries, and all of the serious ones, could have been prevented without compromising the efficacy of the procedure.[5] It should also be noted that the potential for stochastic effects from interventional procedures exists given the increasing number of young and middle-aged patients.

In view of the above concerns about the potential for skin damage, the importance of developing local clinical protocols for each type of interventional procedure is recommended. The protocol should include a statement on the *nominal* values for the technical factors, such as fluoroscopy times, air kerma rates, and resulting cumulative dose at each skin site exposed. A medical physicist expert may be required to assist in establishing this information. The technical factors should relate to the equipment installed at the facility. It must be understood that each clinical case may vary considerably, and the protocol should act only as a baseline for the procedure. The recent move by the International Electrotechnical Commission[37] requiring new interventional equipment to provide real-time indication of dose rates and of the cumulative dose at the end of the procedure, at a position indicative of the patient's skin entrance, is most welcome in terms of patient protection. It will help to alert operators in real time of the potential for skin damage before the threshold for such effects is reached and will certainly highlight to clinical staff those patients who may require monitoring for potential skin damage after the event.

A number of explicit precautions should be implemented in addition to those outlined in Section 6.2.4, and the following advice is offered:

- Options for positioning the patient or altering the x-ray field or other means to alter the beam angulation when the procedure is unexpectedly long should be considered. This will achieve a degree of skin sharing and ensure that the same area of skin is not continuously in the direct x-ray field. For example, in neuroradiological procedures consider using both lateral views rather than just one.

- Take advantage of the equipment options that utilize significant x-ray beam filtration as part of their dose-saving protocols. Significant amounts of copper filtration (say, 0.5 mm) or even k-edge filtration may be employed depending on the age and manufacturer of the equipment. As previously noted in Section 6.2.1.2, thorough continual training and familiarity with the specific equipment is required for the operator to take full advantage of these features. The training should include an appreciation of the magnitude of the skin doses delivered to patients from procedures they undertake on the equipment they use and an awareness of their potential to cause injury.

The ICRP[5] in reviewing the cause of a number of serious radiation injuries to patients highlighted this issue. To quote:

> In many of these cases, it appears certain that the physicians performing the procedures had no awareness or appreciation that the absorbed dose to the skin was approaching or exceeding levels sufficient to cause inflammatory and cell-killing effects.

6.2.6 Specific Protection Issues Relating to Computed Tomography

Technical and clinical developments in computed tomography (CT) have not led in general to reductions in patient dose per examination, in contrast to the trend in general radiology. Coupled with an increased use of CT in diagnosis in most developed countries,[38,39] there are increased concerns about the magnitude of the doses that arise from CT examinations and the potential risks that these imply. The latest mortality data about the Japanese bomb survivors[40] are consistent with there being a risk of cancer induction at doses typical of CT examinations. This is particularly true of CT examinations of pediatric patients, who may also be at greater risk from stochastic effects than the general population.[41] Further, repetitive CT examinations (e.g., multiphasic contrast) have the potential to result in absorbed doses in tissues that may approach or even exceed the threshold for deterministic effects.

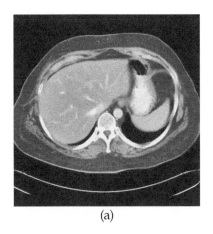
(a)

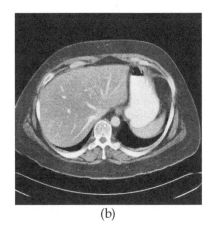
(b)

FIGURE 6.7
Optimization of CT scanning parameters. The two images are of the same patient taken a few weeks apart, matched anatomically as best as possible, but acquired with substantially different technique factors at 120 kVp with the same reconstruction algorithm and slice width of 8 mm. In (*a*) the effective mAs is 157 while in (*b*) it is 94. Noise is more obvious in (*b*), but the image remains diagnostic at a dose reduction of 40% compared with the manufacturer's default protocol.

Multidetector CT scanners offer a number of clinical advantages but, because of a combination of their unique design characteristics and superior scanning speed, are particularly prone to delivering high patient doses[38,42-45] unless technical factors are carefully selected by the operator. Operators should be mindful that the default protocols provided by manufacturers of multidetector CT scanners can often be modified to achieve substantial dose reductions without loss of diagnostic integrity by tailoring the technical parameters used in an examination,[42,46-48] as illustrated in Figure 6.7.

It is clear then that protocols should be developed for all common CT procedures. Further, in the absence of x-ray tube current modulation technology, discussed below, if patient dose and image quality are to be jointly optimized, the operator of a CT scanner should tailor the technical factors of the examination (kVp, effective tube current-time product (mAs), nominal collimated x-ray beam width, pitch, volume of patient scanned) to the individual patient anatomy and the diagnostic information being sought.[38,43,44,49] For a particular patient, all other factors kept constant, the patient effective dose will increase in direct proportion to the mAs and inversely as the pitch. Thus, with single-slice scanners it has been good practice to choose the highest value for the pitch and the lowest value of the mAs consistent with obtaining the required clinical diagnosis. Since a pitch value of less than one is analogous to overlapping scanning in sequential mode, pitch values have usually been chosen in the range of one to two, and only in exceptional circumstances have they been chosen as less than one. With multidetector scanners, some manufacturers have tied the selection of mAs and pitch together so that the ratio of the mAs to pitch (called the effective mAs) remains constant when the pitch is altered. Under these circumstances, changing the pitch has minimal impact on patient dose, and pitch values of less than one may be safely used.

As an example of the customization that may be achieved, Boone et al.[50] have produced patient size-dependent technique charts for one model of a multidetector scanner. Their work, based on phantom simulations, suggests that for pediatric, abdominal CT scans, the mAs may be reduced to less than 5% of the value used for a typical adult, while maintaining the same image quality (contrast-to-noise ratio kept constant). The resulting effective dose reduction is almost as impressive. This work also suggests similar optimization is possible when performing head CT scans on children. Certainly, the need for vigilance in establishing CT scan protocols for pediatric patients has been highlighted.[51]

One of the key advances that may lead to significant dose reduction is the concept of anatomy-dependent, attenuation-based methods of x-ray tube current modulation that has been introduced on newer scanners.[43,52,53] This is often described as a form of AEC, and an excellent explanation of how the manufacturers have implemented this technology has been provided by Keats[53] and others.[55,56] The reader should be aware that the use of AEC does not, in itself, guarantee dose reductions because decisions must be made, based on clinical need, as to what constitutes acceptable image quality. This decision is usually made on the basis of achieving an acceptable level of image noise.

Recommendations concerning achievable standards of good practice in CT have been developed by the European Commission in the form of quality criteria.[44,49] These documents provide an operational framework for radiological protection initiatives in which technical parameters for image quality are considered in relation to patient dose. Diagnostic and dose requirements for CT are specified in terms of the quality criteria considered necessary to produce images of standard quality for a particular anatomical region. The subjective image criteria include anatomical criteria that relate to the visualization or critical reproduction of anatomical features. Criteria concerning patient dose are given in terms of DRLs associated with the examination technique used for standard-sized patients. Quality criteria have been developed for most CT examinations, together with examples of technique parameters influencing the dose.

Specific advice on optimizing multidetector CT protocols[47,48] is summarized below.

- Restrict the scanned volume to the minimum and scan in one large block rather than in multiple smaller contiguous blocks, although there are some possible exceptions to this rule. This minimizes the impact of *overranging* or *overscanning*. Overranging is a necessary consequence of the fact that a CT scanner operated in helical mode would have an incomplete data set from which to reconstruct both the first and last slices of interest if extra rotations were not performed. Typically, an extra rotation is required at the beginning and end of each helical acquisition so that the extent of the effect will depend on the pitch and the x-ray beam collimation.[57–59] Overranging is of particular concern when attempting to avoid irradiating the male gonads in a scan of the lower abdomen/pelvis or the lens of the eye in a helical head scan. There are at least two possible exceptions to the general advice outlined above. First, image noise may be tolerated more readily in one part of the body than another. For example, a higher level of noise may be accepted in the chest than in the abdomen (or vice versa). In these circumstances it would seem sensible to scan these body parts separately, and any dose penalty from overranging would be more than offset by the decreased dose in one or another of the two smaller blocks. A second exception arises when considering scanning of the head and neck, or head and trunk in one block. Many scanners are designed to operate with different beam filtration for the head versus the abdomen, and this filtration cannot be altered in mid-scan. The radiation output per mAs is typically 25% and may be as much as 40% higher in head mode versus body mode.[60] This would mean that the neck or trunk may receive a significantly higher dose than either anticipated or necessary. Thus, helical scanning of the head and neck, or head and trunk should be carried out in two discrete blocks in these instances. Ultimately, the decision about the scanning protocol needs to be made on the basis of the particular technology utilized.

- Avoid transferring scan protocols applicable to one scanner to another without due consideration to differences in scanner filtration and geometry. Even scanners from the same manufacturer differ significantly in their detector type (efficiency), geometry, and filtration.

- Use the widest beam collimation consistent with clinical requirements (e.g., 16 × 1.5 mm rather than 16 × 0.75 mm). A feature of all multidetector scanners with more than two detector rows, whether used in sequential or helical mode, is that they irradiate a significantly larger slice of tissue than might be expected on the basis of the detector size. The problem is referred to as *overbeaming*, but the root cause is the finite size of the focal spot. By using the widest possible collimation, the impact of overbeaming on dose is minimized.

- Consider using lower tube potential for CT angiography given the increased subject contrast provided by the iodine contrast. To be effective, this strategy may mean tolerating a modest increase in image noise.

- Consider using lower tube potentials and certainly considerably lower mAs values with children.

- Keep the effective mAs, which is defined as the quotient of the mAs divided by the pitch, as low as clinically indicated. Patient dose scales directly as the effective mAs, all other factors remaining the same.

- Use anatomy-dependent, attenuation-based methods of x-ray tube current modulation (AEC) with an appropriately selected reference effective mAs or noise index. For optimum performance with this technology the scan projection radiograph (localizer) must be acquired over the full length of the patient that is of clinical relevance using the same kVp that will be subsequently utilized for the volume acquisition.

- Minimize the use of multiphase examinations. This is a fundamental decision relating to justification and optimization, as discussed in Section 6.2.1.2.

- Use sequential as opposed to helical techniques for routine head scans unless clinical indications suggest otherwise. As noted previously, overranging will inevitably lead to irradiation of the lens of the eye if helical mode is employed.

- Avoid the use of CT-perfusion studies (continuous or repeated scanning to follow the time course of injected contrast agent), and if they are undertaken, ensure that an established, optimized protocol is followed. In particular, the scan duration should be limited to the absolute minimum required, as hair loss can occur in patients undergoing this procedure.[61]

In addition to CT-perfusion studies, there are other novel CT applications ranging from CT fluoroscopy to functional and four-dimensional CT. The latter can use a gating signal to select retrospectively CT projections belonging to a particular phase of, for example, the breathing or cardiac cycle. In order to have a sufficient number of projections in all phases, these scans typically use significant oversampling, for example, by using a very small pitch in helical CT. The resulting scanning times are usually many times longer than in conventional CT scanning and could therefore also result in much larger doses received by the patient if the tube current is not reduced. The operator should also be aware that tube current modulation, as mentioned above, may not be available with these modern CT techniques.

Attempts at shielding relatively radiosensitive tissues such as the breast and eyes have been discussed in the literature. During CT scanning of the chest and upper abdomen, substantial breast dose reductions have been demonstrated without compromising diagnostic image quality, by using thin bismuth breast shields raised above the surface of the chest,[62–64] and this approach has some merit. Likewise, bismuth eye shields may also be useful in minimizing the dose to the lens of the eye during head CT examinations,[65] although other investigators[64] have shown that appropriately tilting the gantry offers better dose reduction possibilities. In any event, if tube current modulation technology is employed, as it should be whenever possible, the use of body part shielding is likely to be counterproductive, as the controlling software senses the increased attenuation presented by the shielding when it is directly in the primary beam and increases the instantaneous tube current to ensure that a reasonable transmitted x-ray intensity is maintained at the detectors. Thus, the use of breast and eye shielding would seem contraindicated in these circumstances.

It should be apparent from the practical advice offered above that training is a key component of the optimization process in CT scanning. Any training must relate to the site-specific CT scanner and should address the impact of the scanning parameters on patient dose and image quality as part of the optimization process. Operators need to be able to tailor these parameters to fit the need of the specific examination on an individual patient basis. Operators should also be able to interpret the significance of the dose index $CTDI_w$ (or its equivalent*), which must be displayed on the operator's console of new CT scanners before irradiation, and to understand how the scanner's AEC operates.

---

* The CTDI and related parameters are defined in Appendix A.

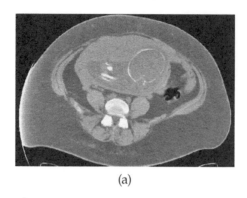

(a)

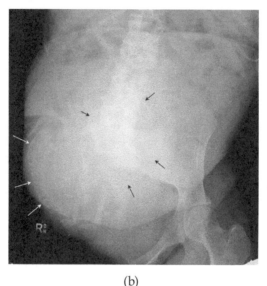

(b)

FIGURE 6.8
In (*a*) a CT scan of an obese 130 kg woman presenting with back pain is shown. She had
responded in the negative when asked about the possibility of pregnancy. The cause of her
back pain is plainly obvious. Estimated fetal age of 26 weeks and fetal equivalent dose estimate
of less than 30 mSv. In (*b*) an abdominal plain x-ray of a known pregnant women clearly shows
the fetal head (black arrows) and spine (white arrows). Fetal dose estimate less than 1 mSv.

6.2.7   Specific Protection Issues Relating to High-Risk Groups

6.2.7.1 Pregnant or Potentially Pregnant Women

Additional precautions[25,66,67] should be taken when irradiation of pregnant or
potentially pregnant women is contemplated. The main aim is to avoid unin-
tentional exposure of an unborn fetus (see Figure 6.8a), or minimize the dose
to the fetus when it cannot be avoided (see Figure 6.8b). Accordingly, it should
be stressed that radiation exposure of the abdomen and pelvis of women

TABLE 6.1

Approximate Fetal Equivalent Doses (mSv) arising from Common Radiological
Examinations of Pregnant Patients

Radiographic/Fluoroscopic Examination	1st Trimester	3rd Trimester
Skull	<0.01	<0.01
Chest	<0.01	<0.01
Cervical spine	<0.01	<0.01
Thoracic spine	<0.01	<0.01
Lumbar spine	2	6
Abdomen	1.5	2.5
Pelvis	1	2
Intravenous pyleogram (IVP)	2	10
Extremities	<0.01	<0.01
Mammography	<0.01	<0.01
Barium meal	1	6
Barium enema	7	25

Note: Values based on data from Sharp et al.[99] and simulations using the PCXMC code.[74]

of reproductive capacity should be kept to a minimum. During pregnancy, radiation exposure to these regions should only occur if the procedure cannot be postponed because of the urgent nature of the investigation.

Thus, it is important to establish the pregnancy status of women of childbearing age whenever exposure of the abdomen and pelvis is contemplated. From an administrative perspective, it may be easier to extend this requirement to all radiological examinations so that all women of reproductive age should be asked about the possibility of being pregnant in a manner that is not seen as intrusive by the patient. Multilingual signs, posted in changing cubicles, asking the patient to notify the radiographer about a possible pregnancy before the examination begins is one precaution that can be taken that does not detract from the responsibility of the attending clinical staff to inquire about the possibility of pregnancy. Notwithstanding this requirement, it should be clear that plain radiography of areas remote from the lower abdomen may be performed safely without regard to the pregnancy status of the patient. For example, skull, chest, dental, cervical spine, or extremity radiography and mammography can be undertaken with negligible exposure to the fetus at any time during pregnancy provided proper collimation is used (see Table 6.1). CT examinations of the head and neck can probably be undertaken safely, but CT of the thorax and abdomen are best avoided unless pregnancy can be excluded (see Table 6.2).

Clearly, it would be prudent to consider as pregnant any woman of reproductive capacity whose menstrual period is overdue or clearly missed at the time of presenting for radiological examinations. In any event, when doubt

TABLE 6.2

Approximate Fetal Equivalent Doses (mSv) Arising from Common CT
Examinations of Pregnant Patients

CT Examination	1st Trimester	3rd Trimester
Brain, routine	<0.005	<0.005
Neck	<0.005	<0.01
Chest, routine without portal phase	0.1	0.6
Chest with portal phase	1.0	7
Chest, routine with high resolution	0.1	0.6
Chest/abdomen/pelvis	12	13
Abdomen/pelvis, routine	12	12
Abdomen/pelvis, triple phase	15	30
Renal (KUB)	9	9
Thoracic spine	0.2	1.0
Lumbar spine	9	23
Angiography, pulmonary	0.1	0.4
Angiography, aortic	11	12
Pelvimetry	—	0.2

Note: Values obtained using the ImPACT dose calculator[60] and typical technique factors.[46]

exists about the pregnancy status of an individual woman and moderate or
high doses to the lower abdomen are involved, serum βHCG testing prior to
medical exposure may be considered. If pregnancy is indicated, consideration must be given to the possibility of delaying the procedure at least until
such time as the fetal sensitivity is reduced (ideally post 24 weeks and certainly post 15 weeks).[66,68] However, delaying a study may be counterproductive, and if in consultation with the referring physician it is decided that the
risk of not making a necessary diagnosis is greater than that of irradiating
the fetus, then the examination should be performed.

If pregnancy is established, it is recommended that, wherever possible,
ultrasound or magnetic resonance imaging (MRI) examinations be substituted for radiological procedures. Ionizing radiation should only be used
when diagnosis cannot be confirmed by other techniques or to provide additional information when fetal abnormality has been demonstrated.

6.2.7.2 Pediatrics

The radiation exposure of growing children should be minimized, given
some evidence[2] to suggest that their tissues are more radiosensitive than
those of mature adults, and their greater life expectancy means that any radiation-induced deleterious effects have a greater potential for manifestation.

Radiography of children differs from that of adults in a number of respects. Infants and smaller children are frequently less cooperative than adults, have faster respiration and cardiac rates, and will often not remain still during the procedure. The potential for substantial retake incidence is quite apparent. To counter some of these problems, it is important to gain the confidence of the child through adequate rapport, perhaps by having a resident radiographer specially trained in pediatric radiological methods. Certainly, patience is required. Other options for dose reduction may include:

- Using mechanical means of immobilization (e.g., compression bands, sand bags, and tapes).
- Using very short exposure times in plain film radiography, as these will minimize the effect of motion artifact. An extension of this concept is to use high frame rates (60 f/s), with very short frame length in cardiac procedures, although high frames are in themselves undesirable because of the increased radiation dose implicit in such studies. Certainly, the use of biplanar angiography facilitates the reduction of dose and contrast administration.
- Using anesthesia or sedation in rare instances, such as CT. In rare instances it may be necessary for a carer (parent or guardian) to assist in holding a child, and this should be done in preference to having staff hold the patient.
- Not using grids in radiography and fluoroscopy, as previously noted. Since the amount of scatter is small with all but the largest children, grids are not generally required. The resulting reduction in radiation dose may be a factor of three or more.

6.2.8   Dose Calculations for Patients

6.2.8.1   Basic Principles

For any practice a knowledge of typical patient doses for common procedures is crucial in the optimization process. The entrance skin dose (ESD) has already been noted as being a useful dosimetric parameter in the context of determining compliance with DRLs (see Section 6.2.1.4). Several methods or combinations of techniques are available for the determination of the ESD and organ absorbed doses (see, for example, Heggie et al.[25]). All of them require a degree of calculation with the exception of phantom measurements. These entail the placement of tissue equivalent models in the x-ray beam. They are designed to simulate the way in which a patient or part of the patient absorbs and scatters ionizing radiation, so that the measurements reflect accurately the anatomical dose distribution occurring clinically. The degree to which the phantom mimics the human body varies with the sophistication and expense of the particular phantom chosen. One such phantom available is the so-called Rando Phantom, which consists of a series of 25 mm thick transverse slices each containing a number of plugs that can be replaced by thermoluminescent (TLD) capsules or pellets (see Chapter 5).

To determine the absorbed dose to organs during a diagnostic procedure, the phantom is placed in the primary beam in exactly the same anatomical position as would be the patient. The x-ray unit is set to the same choice of field size, beam filtration, kVp, and mAs and the exposure initiated. The TLD are removed, evaluated, and the dose distribution obtained. Usually, one is concerned with absorbed dose to the uterus or ovaries or some other important organ, but occasionally the dose distribution could be used to obtain the effective dose.

Of the indirect methods employed, that using skin absorbed dose measurements with TLD has proven popular since the TLD does not intrude into the diagnostic interpretation of the radiograph. This technique is used primarily where surveys of patient exposures are conducted. The TLD pellet or sachet is taped to the patient's skin in the center of the primary beam before the radiograph is taken. Following irradiation, the TLD is read out in the usual manner to obtain the skin absorbed dose. Calibration of the TLD is achieved by exposing it to a known air kerma and correcting for the difference in mass energy absorption coefficients between air and tissue. A knowledge of the field size, beam filtration, and kVp then allows a calculation of the absorbed dose at depth in tissue by reference to tables of percentage depth doses (PDDs).[69]

A similar and more basic approach, and one that can also be applied to determine patient exposure after the event, involves the measurement of the x-ray machine radiation output in air, again reproducing the technique factors employed in the patient procedure. The measurement of machine output is achieved using accurately calibrated ionization chambers or solid-state detectors. The sequence of events required in the task of obtaining the absorbed dose to tissue at depth is:

- Measure the output (K) of the x-ray unit at a convenient reference point, at distance, d, from the x-ray tube (see Figure 6.9).
- Calculate output of the unit at the source–skin distance (SSD) using the inverse square law. This may be denoted as the entrance surface air kerma (ESAK).
- Following the recommendations of the IPSM,[70] calculate the entrance surface dose (ESD) by correcting for backscatter using tabulated backscatter factors.[71–73] Observe that the back scatter factor (BSF) depends on kVp, filtration (HVL), and the irradiated field size (A) at the patient surface entrance.
- Calculate the absorbed dose to tissue at depth, t, using tables of PDD, which depend on kVp, filtration (HVL), A, SSD, and t.[69]

The application of this recipe may be illustrated (see box) in the estimation of the fetal dose following an abdominal x-ray of a pregnant woman. The technical details of the exposure are 75 kVp, 30 mAs, K = 60 µGy/mAs at 1 m, HVL

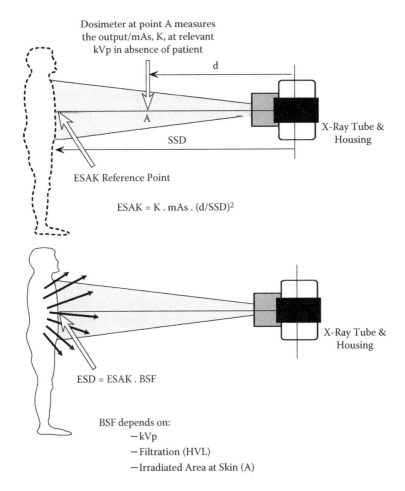

FIGURE 6.9
Schematic diagram illustrating estimation of the ESD. In the upper panel entrance surface air kerma ESAK can be estimated from the output of the x-ray tube. In the lower panel the ESD is estimated from the ESAK and the back scatter factor (BSF).

= 2.7 mm Al, SSD = 75 cm, A = 900 cm^2, and depth of fetus below skin entrance = 8 cm. From the literature one obtains the BSF = 1.35 and PDD = 23.

$$
\begin{aligned}
D_{fetus} \quad &= ESD \times PDD/100 \\
&= ESAK \times BSF \times PDD/100 \\
&= K \times mAs \times d^2/SSD^2 \times BSF \times PDD/100 \\
&= 60 \times 10^{-3} \times 30 \times 100^2/75^2 \times 1.35 \times 23/100 \\
&= 0.99 \text{ mGy}
\end{aligned}
$$

The above process, while fundamentally sound, suffers because its application is very labor intensive, is almost impossible to apply in the context of complicated examinations such as fluoroscopy and CT, and is of little use in determining either the stochastic risk (effective dose) or the absorbed dose to a tissue at risk that may not be in the primary beam. Fortunately, recently software-based dose calculators capable of running on a personal computer have emerged.

6.2.8.2 Dose Calculators

A dose calculator applies a purely mathematical (stylized) model of the transport of photons through the body using the Monte Carlo methodology. The interaction of an individual photon with the tissue (coherent scattering, Compton scattering, and photoelectric effect) is known to be a purely statistical or random phenomenon depending on both the tissue composition and the photon energy. With the Monte Carlo method an individual photon is traced from its point of production in the x-ray tube through a body-simulating mathematical phantom,* in which its energy may or may not be totally dissipated. If the x-ray still possesses some energy, it will ultimately exit from that phantom and may interact in the image receptor. Indeed, it is possible to follow many hundreds of thousands or more of such photons, tracing their histories as they interact with various tissue types and noting any local energy deposition that occurs. Ultimately, a reliable energy distribution or measure of the tissue absorbed doses within the body be obtained. However, the success of the model depends on how well the assumed computer geometrical model simulates the body. The mathematical phantom must define the organ boundaries correctly and unambiguously, and given the wide variation in anatomy encountered in radiology, this is unlikely to be the case for any given individual. Nevertheless, this technique remains one of the most robust means of determining typical patient absorbed doses, and a recent application of this methodology[74] provides estimates of organ and effective dose. The calculation is in three parts (see Figure 6.10): entry of examination data, simulation of x-ray transport through patient, and calculation of doses based on the actual technique factors utilized. While there can be no denying the usefulness of this software, its limitations must be accepted: specifically, it uses a geometrical model of the patient and the actual organs cannot be expected to be located correctly anatomically or to be necessarily of the correct size or shape as those of a real patient.

With regard to the dosimetry of CT examinations there are two aspects that should be considered. While the dose length product (DLP) is recommended for comparison with DRL, and is now a required displayed parameter postexposure on the console of newer CT scanners,[75] it is equally important that a means to calculate organ and effective doses be available. Fortunately, there are at least

---

* Recently, voxel models of humans based on scans of real individuals of both sexes and various ages have been created, and these promise to refine the dosimetry process further.

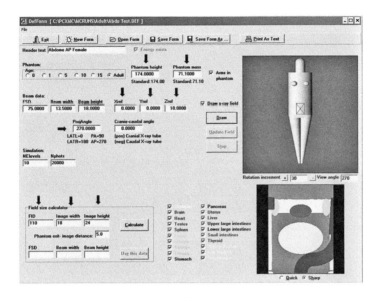

(a)

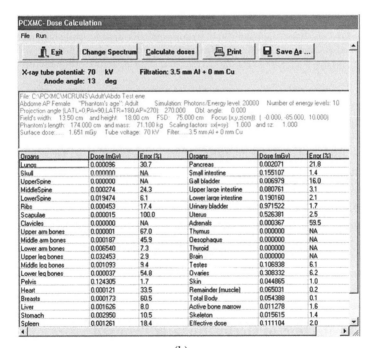

(b)

FIGURE 6.10

Use of PCXMC[74] to calculate organ and effective dose from plain abdomen examination. The *examination data* and *calculation of doses* pages are illustrated.

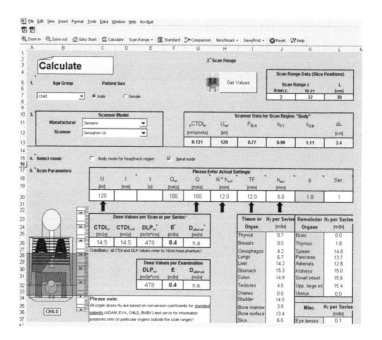

FIGURE 6.11

Use of the CT-Expo dose calculator[77] to determine effective dose for child abdominal examination. Scanned volume is illustrated at bottom left, key entry data items are centrally located (see arrows), organ doses are shown at bottom right, and key dose indices are shown at lower center.

three dose calculators commercially available[60,76,77] and all are widely used. An example of the use of CT-Expo to calculate the effective dose for a child undergoing an abdominal CT examination is illustrated in Figure 6.11.

Again, the user must be acutely aware of the limitations of such programs, as the Monte Carlo calculations implicit in them use a stylized phantom similar to that utilized in PCXMC. Further, some of the assumptions inherent in these calculators about the spectra emanating from the CT x-ray tubes and the scanner geometry may not be strictly applicable to newer scanners in spite of the best efforts of the authors to match them.

6.3 Protection of Occupationally Exposed Individuals and the Public

6.3.1 Recognition and Avoidance of Hazards

It is important to recognize that the hazard to the occupationally exposed individual in radiology arises from three sources (Figure 6.12). These are the

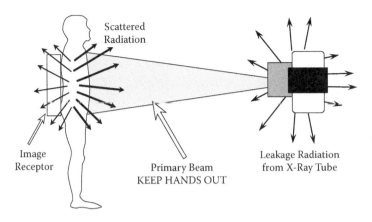

FIGURE 6.12
Recognize the exposure hazards: leakage radiation, scatter radiation, and primary radiation.

primary x-ray beam, x-ray tube leakage radiation (see Section 6.1.3), and, most important of all, scatter from the patient. These latter two sources of hazard are collectively called secondary radiation. It should be rare for clinical staff to be exposed to the primary beam. However, in those instances where uncooperative patients are involved it may be necessary to hold or restrain them during a procedure, and restraining devices should be utilized whenever possible. In the case of children it is invariably good practice to allow a carer to assist. In all instances leaded gloves should be worn if there is any chance of the hands being in the primary beam.

Personnel required to remain in an x-ray room during an exposure or in theatre during screening should remember one of the basic tenets of radiation protection: maximize the distance between the source(s) of the hazard and themselves. Doubling the distance from the patient and x-ray source to oneself will reduce any occupational exposure by approximately a factor of four. This is confirmed by published figures [78] suggesting that for most common radiographic examinations the level of scatter reduces to less than 0.5% and 0.1% of the patient ESD at 1 and 2 m from the patient, respectively. Certainly, all potentially exposed individuals should wear full wraparound lead aprons (not backless aprons) to minimize the impact of leakage and scatter radiation. Note that, depending on their lead equivalence, aprons attenuate typically about 90–95% of the radiation[79] but never all of it. In any event, the number of personnel in the immediate vicinity of the patient should be kept to a minimum and some warning given by the radiographer or operator that an exposure is impending. Ideally, personnel should retreat to the safety of the operator's console, if one exists, or leave the room during an exposure as, normally, the design of a radiology suite is such that either adequate shielding or distance ensures that any individual external to the room or behind the console will receive an inconsequential radiation exposure.

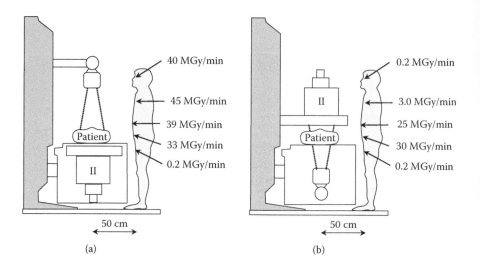

FIGURE 6.13
Typical skin absorbed dose rates near fixed fluoroscopic equipment in the absence of protective aprons or drapes. (*a*) Overtable x-ray tube. Note elevated dose rates to trunk and head. (*b*) Undertable x-ray tube with relatively high dose rates only in region of lower trunk.

Conventional fluoroscopic installations present no great hazard if the above precautions are adhered to. However, it is worth noting that the configuration employing an overtable x-ray tube and an undertable image intensifier is intrinsically less safe than the more traditional configuration (see Figure 6.13). Primarily this is because leakage and scatter radiation are significantly reduced by the image-intensifier housing and table in the case of the latter configuration. Accordingly, the use of the above table x-ray tube configuration should be discouraged unless it can be operated remotely.

6.3.2   Specific Issues for Cardiology and Interventional Radiology

The practice of using C-arms in cardiac and interventional angiography suites does increase the potential radiation hazard compared with using conventional fixed fluoroscopic equipment. There are two reasons for this. First, there are frequently no lead drapes or built-in shielding to minimize scatter and leakage at the operator's position. Second, as angioplasty, stenting, and radio frequency ablation procedures become more common, there has been a dramatic increase in fluoroscopic screening times. It has been reported that substantial occupational exposure may arise during cardiology and interventional procedures as a result of inappropriate equipment and inadequate personnel protection.[80] As previously noted, the major radiation hazard is scatter radiation emanating from the patient, and in general, occupational doses will scale with patient doses so that occupational doses can be lowered

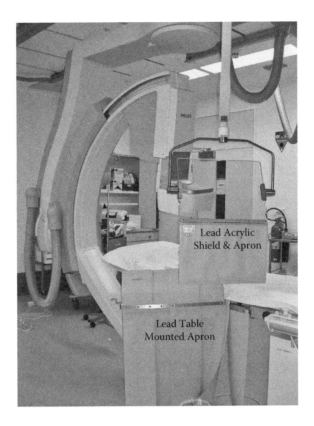

FIGURE 6.14
Overhead mounted lead acrylic viewing window with suspended lead drapes and table-mounted lead drapes offers improved protection in angiography. The x-ray tube is obscured by the table-mounted drapes, which effectively minimizes the leakage radiation reaching the operator. The operator controls for adjusting gantry position; FOV, collimators, and so forth, are at bottom right.

by reducing unnecessary patient dose and also by the procurement and use of appropriate equipment, including shielding devices.[5,81,82]

It is also quite possible for the lens of the eye and the thyroid to receive substantial doses, perhaps even as high as the recommended equivalent dose limit in the case of the former, in the absence of steps to limit exposure to these tissues. Indeed, several cases of lens injuries caused by occupational exposure have been reported,[81] and a recent survey of interventional radiologists reported a significant percentage had clinically noticeable opacities.[83] Thus, it is recommended that thyroid shields and lead glasses be worn by cardiologists and interventional radiologists. The use of drop-down lead acrylic viewing windows is generally acknowledged as offering better eye protection than the wearing of lead glasses, and the use of table-mounted lead aprons is also highly recommended (see Figure 6.14).

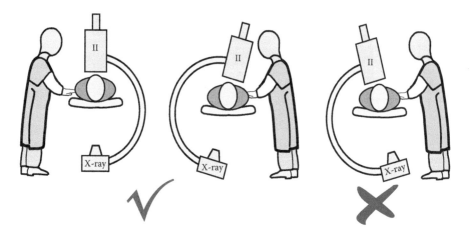

FIGURE 6.15
Preferred geometries from the occupational perspective when using C-arm imaging equipment are shown schematically at left and middle. The orientation on the right suffers from the operator being subjected to more leakage and scatter radiation.

While some general recommendations about good geometry have already been made in the context of patient dose reduction (see Section 6.2.4), some preferred orientations of C-arm x-ray equipment are indicated in Figure 6.15. For lateral and oblique projections the operator should stand on the side of the patient where the imaging device is located.

In some abdominal interventional procedures, radiologists may find it necessary to have their hands in the beam for some part of the procedure. Although this is never a recommended practice, in such circumstances it is vitally important that the hands be on the exit beam side of the patient. For example, if the x-ray tube is positioned under the patient, then the hands manipulating any catheters should be on top of the patient. Failure to adhere to these basic protection rules may lead to severe radiation-induced dermatitis on the hands.[23] The use of forceps may aid in reducing the frequency of such occurrences, and as a final resort, the wearing of leaded gloves may need to be considered, although issues of dexterity and sterility may take precedence. For obvious reasons it is recommended that interventional radiologists wear finger TLDs on their dominant hand to monitor extremity exposures.

6.3.3 Specific Issues for CT Fluoroscopy (Finger Doses)

The issue of radiation protection during CT fluoroscopy needs to be considered carefully since CT fluoroscopy has the potential to result in high doses to the hands of the radiologist performing these procedures. In fact, the dose limits for the extremities may easily be exceeded for a realistic case load.[84]

The recent implementation by some manufacturers of technology that allows the x-ray tube to be switched off as it rotates above the patient represents a commendable step forward in terms of reducing dose to the operator.

Since the wearing of leaded gloves may result in a loss of dexterity, this option may be untenable in some CT fluoroscopy procedures. Accordingly, radiologists should consider using specially designed forceps or needle holders to aid in dose minimization to their fingers.[84]

One practical measure that may be implemented is the use of lead drapes placed approximately 2 cm caudal to the scan plane, as significant dose reductions to both the hands and abdomen of radiologists have been demonstrated.[85] It is highly recommended that this dose reduction technique be implemented during CT fluoroscopy. This in no way absolves the radiologist from the need to wear protective clothing.

6.3.4 Personal Protective Equipment

Lead aprons, thyroid shields, and other personnel protective devices should meet minimum design criteria as outlined in relevant standards (see e.g., references 86 and 87). Although lead aprons must be of at least 0.25 mm lead equivalence, in practice, their thickness should be selected with due consideration given the type of workload being undertaken. Individuals continually involved in interventional radiology should wear aprons of at least 0.35 mm lead equivalence, if not 0.5 mm lead equivalence. Preferred designs are those comprising a separate vest and skirt that wrap around fully, as open-back designs are not recommended. All personnel protective clothing should be examined under fluoroscopy at least annually to confirm the integrity of the protection.

6.4   Shielding Issues

6.4.1   Basic Concepts

The fundamental consideration with any shielding design is to achieve a safe environment for designated occupationally exposed workers and members of the public, including other employees. For many years many shielding designs were based on the principles outlined by the National Council on Radiation Protection and Measurements in NCRP Report 49,[88] and those principles remain valid to this day. However, that report suffered from a number of deficiencies in that the suggested methodology to be followed adopted a number of unduly conservative assumptions that when applied rigorously resulted in massive overestimates of shielding requirements and, by implication, increased costs for radiological practices. For example, in general radiography no allowance was made to account for self-attenuation by the patient, and intervening bucky grid, image receptor, or physical support for the last item. Further, the treatment of leakage was inadequate in that a rather artificial "add an HVL" rule was adopted to account for the fact that leakage radiation was in general of higher beam quality than scatter radiation. It has been demonstrated by Simpkin and Dixon[89] and noted by others[90] that the NCRP Report 49 methodology for handling leakage radiation may lead to overly conservative estimates of the level of transmitted air kerma through a barrier by factors as high as eight thousand. Even the use factors, occupancy factors, and facility workloads suggested in NCRP Report 49 are by and large unrealistically high. Fortunately, revisions of this document have finally been produced[91,92] that have addressed many of these shortcomings.

There are some fundamental concepts and definitions that are common to most methods of calculation,[90,92] and these are outlined briefly. First is the notion of a controlled area, which is an area with access limited to members of the radiology practice and where the exposure of personnel to radiation is under the supervision of a designated radiation safety officer. X-ray procedure rooms and control booths are designated controlled areas. All other areas, for radiation protection purposes, would normally be designated as uncontrolled areas and would be subject to the design dose constraint relevant for members of the public.

Jurisdictions do differ on the values used for the design dose constraints. There is fairly universal acceptance of a design dose constraint of 5 mGy* per year for occupationally exposed individuals,[90,92] but differences emerge when it comes to specifying the design dose constraint for members of the public. For example, in the United Kingdom[90] and New Zealand[30] it is taken to be 0.3 mGy per year, and in NCRP Report 147[92] it is taken to be 1 mGy per year. In many countries the value to be used is currently the subject of much debate.

The exposure of individuals is dictated by:

- The amount of radiation produced by the source. When the source of the hazard is primary radiation emanating from an x-ray tube, this may be expressed in terms of the workload (W), which is the total x-ray tube current summed over a specified time frame, usually taken as a week. Thus, for diagnostic and interventional radiology workloads are traditionally expressed in terms of mA·min per week. In NCRP Report 147[92] values of W for many situations may be equated to an unshielded value of the air kerma (see below).

---

* The NCRP calls design dose constraints "shielding design goals" (P) and quotes them in terms of air kerma per year with units of milligray per year. Its argument is based quite correctly on the fact that air kerma can be measured directly, but the effective dose, which is used in specifying design dose constraints by other bodies, depends in a complicated manner on body position and the radiosensitivity of the irradiated organs. For the balance of the discussion, the dose constraint will be discussed in terms of air kerma.

TABLE 6.3

Primary Beam Use Factors (U) for a General Radiography Room (adapted from NCRP Report 147[91])

Barrier	Use Factor (U)
Floor	1.0
Chest image receptor (bucky)	1.0
Cross table wall	0.89
Any other wall	0.02

- The distance between the exposed person and the source of the hazard. For all practical purposes the air kerma rate varies approximately as the inverse square of the distance from the source of the hazard. When a barrier is in place, it is usually conservatively assumed that the individual to be protected is at least 0.3 m beyond the barrier.

- The amount of time that an individual spends in the irradiated area. Estimating this requires a knowledge of the workload (W) and the fraction of the time while the x-ray source is on that the individual is in the radiation field. This latter fraction is called the occupancy factor (T). Table 4.1 of NCRP Report 147[92] provides values of suggested occupancy factors that may be used when other site-specific occupancy data are not available.

- The use factor (U), which represents the fraction of the primary radiation (see Section 6.3.1), that is directed toward a given barrier. This depends very much on the type of installation and the barrier concerned. For example, in a dedicated chest room the barrier immediately behind the vertical chest bucky will have a use factor of one and all other barriers will have a use factor of zero. Thus, only secondary radiation (leakage plus scatter radiation) need be considered when designing these other barriers. By contrast, the use factor for fluoroscopy and CT installations will usually be zero for all barriers because the primary beam is totally attenuated by the image receptor. For general rooms with a wall bucky, the values presented in Table 6.3 may be used for U.

- The amount of shielding between the individual and the x-ray source. The extent to which this shielding is effective may be judged on the basis of the broad-beam transmission of the barrier, B(x), which may be defined loosely as the ratio of the air kerma with the barrier of thickness x in place to the air kerma at the same point without the barrier. Archer et al.[93] have provided a robust empirical mathematical model that describes the attenuating characteristics of most shielding materials adequately. Subsequent work by Simpkin and others[94–98] has been used to produce the tabulated and graphical values of B(x) for primary and secondary barriers presented in Appendices

A, B, and C of NCRP Report 147,[92] and the reader is referred to that source for specific values relevant to a particular scenario.

Using the nomenclature defined above, the object of the shielding calculation is to determine the barrier thickness that is sufficient to provide protection to a level below the weekly design constraint, P. Thus, we require that

$$B(x) \le (P/T) \, d^2/(U \, N \, K^1) \tag{6.1}$$

where d is the distance from the radiation source (the x-ray tube focus for primary and leakage radiation and the patient entrance surface for scatter radiation) to the individual beyond the barrier, K^1 is the average unshielded air kerma per patient at 1 m from the source, and N is the expected number of patients examined in the room each week. For a secondary barrier calculation the value of U should be one.

6.4.2   General/Fluoroscopy Installations

It is appropriate to illustrate the use of the above method in a few simple examples.

Example 6.1

In a dedicated chest room imaging forty patients per day for each of 5 days, a bucky is placed against a wall abutting a room used for office work. Both U and T should be assumed to be one, and P = 0.02 mGy per week. The x-ray tube focus is 2.1 m from the existing plasterboard partition wall of thickness 150 mm. Table 4.5 of NCRP Report 147[92] suggests that K^1 = 2.3 mGy/patient at 1 m. The required barrier transmission is given by

$$B(x) \le (P/T) \, d^2/(U \, N \, K^1) = 0.02/1 \times (2.1 + 0.15 + 0.3)^2/(1 \times 40 \times 5 \times 2.3) = 0.00028$$

The graphical data in Figure B2 of NCRP Report 147[92] suggest 2.8 mm of lead would suffice. However, this takes no account of preshielding provided by the image receptor and its supporting structures. Table 4.6 of that same reference suggests that the attenuation provided by the image receptor and associated supporting structures is equivalent to 0.85 mm of lead. Thus, the net lead equivalence for the required barrier shielding is 2 mm.

Example 6.2

A fluoroscopy installation examines eight patients per day for each of 5 days using the undertable x-ray tube. The radiographer's control booth has a partition, thickness 150 mm, 2 m from the x-ray tube and the patient, which should be classified as a secondary barrier. Again, both U and T should be assumed to be one, but P = 0.1 mGy per week, since the radiographers may be classified as occupationally exposed. Table 4.7 of NCRP Report 147[92] suggests that $K^1 = 0.32$ mGy/patient at 1 m. The required barrier transmission is given by

$$B(x) \leq (P/T) \, d^2/(U \, N \, K^1) = 0.1/1 \times (2.0 + 0.15 + 0.3)^2/(1 \times 8 \times 5 \times 0.32) = 0.047$$

which, from Figure C2 in Appendix C of NCRP Report 147,[92] suggests that a lead thickness of 0.5 mm would be more than adequate.

Example 6.3

The situation in example 6.2 is revisited, but this time an existing barrier, 1 m from the x-ray tube and patient, is made of concrete of thickness 150 mm abutting a corridor with pedestrian traffic. Again, U should be assumed equal to one, but a sensible value for T is 0.2 and P = 0.02 mGy per week. The required barrier transmission is given by

$$B(x) \leq (P/T) \, d^2/(U \, N \, K^1) = 0.02/0.2 \times (1.0 + 0.15 + 0.3)^2/(1 \times 8 \times 5 \times 0.32) = 0.026$$

which, from Figure C3 in Appendix C of NCRP Report 147,[92] suggests that a concrete thickness of 50 mm would be more than adequate. Because the existing barrier is already 150 mm of concrete, no further shielding is required.

6.4.3 Mammography Installations

There is fairly general agreement[90,92,98] that mammography units rarely require any additional room shielding other than that provided by the materials usually present as part of the building structures and the operator's protective leaded screen. There are two reasons for this. First, the primary radiation is totally intercepted by the breast support (or the patient on the chest wall margin), and second, the x-ray tube potentials traditionally employed are very low, typically 25 to 35 kVp, with molybdenum or rhodium as the anode material in the x-ray tube. In some circumstances, the weakest link in the design may be the shielding provided by the entrance door to the room if an uncontrolled area with relatively high occupancy (T > 1/8) is on the other side. A further cautionary note is that the recent move toward adopting digital technology in mammography has led to an upward shift in the x-ray tube potentials employed so that the issue of the adequacy of mammographic shielding may require revisiting again, even given the conservative nature of the calculations provided in NCRP Report 147.[92]

6.4.4 Computed Tomography

CT employs relatively narrow collimated x-ray fan beams that are fully intercepted by the detector array as the x-ray tube rotates around the patient. As such, the only concern is that arising from secondary radiation. However, the operating tube potential is typically 80 to 140 kVp, and the workload is much higher than that encountered in general radiography or fluoroscopy, so that particular attention needs to be paid to the attenuation requirements of ceilings and floors.

NCRP Report 147[92] discusses several methods for calculating the shielding requirements of CT scanners, including methods based on the CTDI, the DLP, and a method that utilizes scattered radiation isodose contours. By way of a cautionary note, it is noted that calculations using a workload expressed in mA·min per week are no longer recommended for CT. This is because a multidetector CT scanner typically requires a small fraction of the workload of a single-detector CT scanner for the same clinical coverage, yet the scattered air kerma remains comparable.

An example calculation using the DLP method will be illustrated since modern scanners now display this on the operator's console following the acquisition, and this method is possibly the easiest to implement. The defining equations establishing the scatter air kerma, including a small contribution from leakage radiation at 1 m from the patient, may be written as[92]

$$K^1(\text{head}) = 9 \times 10^{-5} \times DLP \tag{6.2}$$

$$K^1(\text{body}) = 3.6 \times 10^{-4} \times DLP \tag{6.3}$$

where K^1(head) and K^1(body) represent the air kerma at 1 m from the isocenter for head and body scans, respectively and the DLP value relates to the total acquisition. In the absence of a site- or scanner-specific value, indicative DLP values may be taken from Table 5.2 of NCRP Publication 147,[92] although it must be pointed out that use of these values will lead to a very conservative estimate of shielding requirements.

Example 6.4

A new multidetector CT scanner is to be installed and the projected patient load is forty per 8 h day, of which 40% are head scans. Half of all scans will be performed with and without contrast. Based on experience with a like scanner, the DLP for the head and body scans may be taken as 700 and 400 mGy·cm, respectively. The shielding is required for a barrier that is 3 m from the isocenter. The abutting area is an office that has an occupancy, T, of one and a dose constraint, P, of 0.02 mGy per week. From equations (6.2) and (6.3) the unshielded weekly air kerma at 1 m for all the head and body scans, respectively, is given by

$$K^1_{\text{head}} = (5 \times 40 \times 0.4 \times 1.5) \times 9 \times 10^{-5} \times 700 = 7.56 \text{ mGy/week}$$

$$K^1_{\text{body}} = (5 \times 40 \times 0.6 \times 1.5) \times 3.6 \times 10^{-4} \times 400 = 25.92 \text{ mGy/week}$$

The total unshielded weekly air kerma at 1 m, K^1, is the sum of these two values so that the required barrier transmission is

$$B(x) \leq (P/T)\, d^2/(U\, K^1) = 0.02/1.0 \times (3.0 + 0.15 + 0.3)^2/(7.56 + 25.92) = 0.0071$$

The required barrier thickness is then obtained from Figure A2 in Appendix A of NCRP Report 147[92] as 1.4 mm of lead.

6.4.5 Dental Units

The shielding requirements for dental units are trivial by comparison with those for other branches of radiology because the workload is usually low and the radiation doses involved are particularly small.[90,91] Occasionally, because of the compromised design of a dental surgery, a dental x-ray unit may be operated close to a wall, and the explicit circumstances surrounding the use of the unit need to be considered. In any event, the primary beam should always be intercepted by the patient, and further, the primary beam transmission may be ignored with panoramic units and assumed to be very low (<2 μGy per image[90]) for intraoral units. The BIR/IPEM Report[90] using a very conservative design dose constraint of 0.3 mGy per year came to the following conclusions with regard to the use of dental intraoral units:

- No shielding is required if the workload is less than twenty images per week and the distance between the patient and wall is at least 2 m.
- Surgery walls using brick or concrete should provide adequate protection under any circumstance.
- Partition walls with 10 mm of plasterboard on both sides will provide sufficient protection in most circumstances.

For panoramic units the same report concluded that even if the unit was mounted very close (say, 0.7 m) to a wall, more than seventeen examinations would need to be conducted per week before the design dose constraint of 0.3 mGy would be exceeded, even if the occupancy was 100%.

References

1. International Electrotechnical Commission, *Approval and Test Specification—Medical Electrical Equipment*, Parts 1–3, *General Requirements for Safety—Collateral Standard: Requirements for Radiation Protection in Diagnostic X-ray Equipment*, IEC 601-1-3. Geneva, Switzerland: IEC, 1994.

2. International Commission on Radiological Protection, 1990 recommendations of the International Commission on Radiological Protection, ICRP Publication 60, *Annals of the ICRP*, 21 (1–3), 1991.

3. International Atomic Energy Agency, *Radiological Protection for Medical Exposure to Ionizing Radiation*, IAEA Safety Standards Series, No. RS-G-1.5. Vienna, Austria: IAEA, 2002.

4. International Commission on Radiological Protection, Managing patient dose in digital radiography, ICRP Publication 93, *Annals of the ICRP*, 34 (1), 2004.

5. International Commission on Radiological Protection, Avoidance of radiation injuries from medical interventional procedures, ICRP Publication 85, *Annals of the ICRP*, 30 (2), 2000.

6. National Radiological Protection Board, Guidelines on patient dose to promote the optimisation of protection for diagnostic medical exposures, *Documents of the NRPB*, 10 (1), 1999.

7. Dance, D. R., et al., The use of carbon fibre material in radiographic cassettes: Estimation of the dose and contrast advantages, *Br. J. Radiol.*, 70, 383–90, 1997.

8. Kaczmarek, V., et al., Results of a nationwide survey of chest radiography: Comparison with results of a previous study, *Radiology*, 215, 891–96, 2000.

9. Boal, T. J., Cardillo, I., and Einsiedel, P. F., Paediatric doses from diagnostic radiology in Victoria, *Austral. Phys. Eng. Sci. Med.*, 21, 57–67, 1998.

10. International Electrotechnical Commission, *Evaluation and Routine Testing in Medical Imaging Departments*, Part 3.1, *Acceptance Tests-Imaging Performance of X-ray Equipment for Radiographic and Radioscopic Systems*, IEC 61223-3-1. Geneva: IEC, 1999. (Note that this is just one of a series of relevant IEC standards covering acceptance testing.)

11. Institute of Physics and Engineering in Medicine, *Recommended Standards for the Performance Testing of Diagnostic X-ray Imaging Systems*, IPEM Report 77. York, UK: IPEM, 1996.

12. American Association of Physicists in Medicine, *Quality Control in Diagnostic Radiology*, AAPM Report 74. Madison, WI: AAPM, 2002.

13. Institute of Physics and Engineering in Medicine, *Measurement of the Performance Characteristics of Diagnostic X-ray Systems Used in Medicine*, Part III, *Computed Tomography X-ray Scanners*, IPEM Report 32. York, UK: IPEM, 2003.

14. Honea, R., Blado, M. E., and Ma, Y., Is reject analysis necessary after converting to computed radiography? *J. Dig. Imag.*, 15 (Suppl. 1), 41–52, 2002.

15. International Commission on Radiological Protection, Radiological protection and safety in medicine, ICRP Publication 73, *Annals of the ICRP*, 26 (2), 1996.

16. George, J., et al., Patient dose optimization in plain radiography based on standard exposure factors, *Br. J. Radiol.*, 77, 858–63, 2004.

17. International Atomic Energy Agency, *International Basic Safety Standards for Protection against Ionizing Radiation and for the Safety of Radiation Sources*, IAEA Safety Series, No. 115. Vienna, Austria: IAEA, 1996.

18. European Commission, *European Guidelines on Quality Criteria for Diagnostic Radiographic Images in Paediatrics*, EUR 16261EN. Luxembourg: European Commission, 1996.

19. European Commission, *European Guidelines on Quality Criteria for Diagnostic Radiographic Images*, EUR 16260. Luxembourg: European Commission, 1996.

20. Workman, A., and Cowen, A. R., Exposure monitoring in photostimulable phosphor computed radiography, *Rad. Prot. Dosim.*, 43, 135–38, 1992.

21. Strickland, N. H., and Allison, D. J., Implementing a hospital wide PACS: The Hammersmith experience, *Radiology*, 189 (Suppl. 163), 1993.

22. Heggie, J. C. P., and Wilkinson, L. E., Radiation doses from common radiographic procedures: A ten year perspective, *Austral. Phys. Eng. Sci. Med.*, 23, 124–33, 2000.

23. Wagner, L. K., and Archer, B. R., *Minimising Risks from Fluoroscopic X-rays*, 3rd ed. Woodlands, TX: Partners in Radiation Management, 2000.

24. Mahesh, M., The AAPM/RSNA physics tutorial for residents—fluoroscopy: Patient radiation exposure issues, *RadioGraphics*, 21, 1033–45, 2001.

25. Heggie, J. C. P., Liddell, N. A., and Maher, K. P., *Applied Imaging Technology*, 4th ed. Melbourne, Australia: St. Vincent's Hospital Melbourne, 2001, chap. 3.

26. Granger, W. E., Bednarek, D. R., and Rudin, S., Primary beam exposure outside the fluoroscopic field of view, *Med. Phys.*, 24, 703–7, 1997.

27. Marshall, N. W., Chapple, C. L., and Kotre, C. J., Diagnostic reference levels in interventional radiology, *Phys. Med. Biol.*, 45, 3833–46, 2000.

28. Standards Australia and Standards New Zealand, *Approval and Test Specification—Medical Electrical Equipment*, Part 1.3, *General Requirements for Safety— Collateral Standard: Requirements for Radiation Protection in Diagnostic X-ray Equipment*, AS/NZS 3200.1.3, AS/NZS, 1996.

29. U.S. Food and Drug Agency, Department of Health and Human Services, *Performance Standards for Ionizing Radiation Emitting Products*, 21CFR 1020. Rockville, MD: USFDA, 2006.

30. National Radiation Laboratory, *Code of Safe Practice for the Use of X-rays in Medical Diagnosis*, NRL Report C5, Christchurch, NZ: Ministry of Health, 1994.

31. Lloyd, P., et al., The secondary radiation grid: Its effect on fluoroscopic dosearea product during barium enema, *Br. J. Radiol.*, 71, 303–6, 1998.

32. Tapiovaara, M. J., Sandborg, M., and Dance, D. R., A search for improved technique factors in paediatric fluoroscopy, *Phys. Med. Biol.*, 44, 537–59, 1999.

33. Vano, E., et al., Dosimetric and radiation protection considerations based on some cases of patient skin injuries in interventional cardiology, *Br. J. Radiol.*, 71, 510–16, 1998.

34. Koenig, T. R., et al., Skin injuries from fluoroscopically guided procedures. Part 1. Characteristics of radiation injury, *AJR*, 177, 3–12, 2001.

35. Koenig, T. R., Mettler, F. A., and Wagner, L. K., Skin injuries from fluoroscopically guided procedures. Part 2. Review of 73 cases and recommendations for minimizing dose delivered to patient, *AJR*, 177, 13–20, 2001.

36. World Health Organization, *Efficacy and Radiation Safety in Interventional Radiology*. Geneva, Switzerland: WHO, 2000.

37. International Electrotechnical Commission, *Medical Electrical Equipment: Particular Requirements for Safety-X-ray Equipment for Interventional Procedures*, IEC 60601-2-43. Geneva, Switzerland: IEC, 2002.

38. International Commission on Radiological Protection, Managing patient dose in computed tomography, ICRP Publication 87, *Annals of the ICRP*, 30 (4), 2000.

39. United Nations Scientific Committee on the Effects of Atomic Radiation, *Report to the General Assembly*, Annex D, *Medical Radiation Exposures*. New York: United Nations, 2000.

40. Preston, D. L., et al., Studies of mortality of atomic bomb survivors, Report 13, Solid cancer and non-cancer disease mortality: 1950–1997, *Radiat. Res.*, 160, 381–407, 2003.

41. Brenner, D. J., et al., Estimated risk of radiation induced fatal cancer from pediatric CT, *AJR*, 176, 289–96, 2001.
42. International Commission on Radiological Protection, Educational PowerPoint template on managing patient dose in computed tomography, 2001, available on the ICRP website at http://www.icrp.org/educational_area.asp.
43. Nagel, H. D., et al., *Radiation Exposure in Computed Tomography: Fundamentals, Influencing Parameters, Dose Assessment, Optimisation, Scanner Data Terminology.* Hamburg, Germany: CTB Publications, 2002.
44. European Commission, *CT Safety and Efficacy: A Broad Perspective: 2004 CT Quality Criteria*, http://www.msct.info/CT_quality_Criteria.htm), accessed April 24, 2007.
45. International Commission on Radiological Protection, Managing patient dose in multi-detector computed tomography (MDCT), ICRP Publication 102, *Annals of the ICRP*, 37(1), 2007.
46. Heggie, J. C. P., Patient doses in multi-slice CT and the importance of optimisation, *Austral. Phys. Eng. Sci. Med.*, 28, 86–96, 2005.
47. Heggie, J. C. P., Kay, J. K., and Lee, W. K., Importance in optimization of multi-slice computed tomography scan protocols, *Austral. Radiol.*, 50, 278–85, 2006.
48. Kalra, M., et al., Strategies for CT radiation dose optimization, *Radiology*, 230, 619–28, 2004.
49. European Commission, *Quality Criteria for Computed Tomography*, EUR 16262. Luxembourg: European Commission, 1999.
50. Boone, J. M., et al., Dose reduction in pediatric CT: A rational policy, *Radiology*, 228, 352–60, 2003.
51. McLean, D., Malitz, N., and Lewis, S., Survey of effective dose levels from typical paediatric CT protocols, *Austral. Radiol.*, 47, 135–42, 2003.
52. Gies, M., et al., Dose reduction in CT by anatomically adapted tube current modulation. I. Simulation studies, *Med. Phys.*, 26, 2235–47, 1999.
53. Kalender, W. A., Wolf, H., and Suess, C., Dose reduction in CT by anatomically adapted tube current modulation. II. Phantom measurements, *Med. Phys.*, 26, 2248–53, 1999.
54. Keats, N., *CT Scanner Automatic Exposure Control Systems*, MHRA Report 05016. London: MHRA, 2005.
55. McCollough, C. H., Bruesewitz, M. R., and Kofler, J. M., CT dose reduction and dose management tools: Overview of available options, *RadioGraphics*, 26, 503–12, 2006.
56. Kalra, M., et al., Techniques and applications of automatic tube current modulation for CT, *Radiology*, 233, 649–57, 2004.
57. Kalender, W. A., *Computed Tomography: Fundamentals, System Technology and Image Quality Applications*, 2nd ed. Erlangen, Germany: Publicis Corporate, 2005.
58. Tzedakis, A., et al., The effect of overscanning on patient effective dose from multidetector helical computed tomography examinations, *Med. Phys.*, 32, 1621–29, 2005.
59. Van der Molen, A. J., and Geleijns, J., Overranging in multisection CT: Quantitative and relative contribution to dose-comparison of four 16 section CT scanners, *Radiology*, 242, 208–16, 2007.
60. ImPACT, Patient Dose Calculator, Version 0.99x, January 20, 2006. Available from http//www.impactscan.org.

61. Imanishi, Y., et al., Radiation-induced temporary hair loss as a radiation damage only occurring in patients who had the combination of MDCT and DSA, *Eur. Radiol.*, 15, 41–46, 2005.
62. Hopper, K. D., et al., The breast: In-plane x-ray protection during diagnostic thoracic CT—shielding with bismuth radioprotective garments, *Radiology*, 205, 853 58, 1997.
63. Fricke, B. L., et al., In-plane bismuth breast shields for pediatric CT: Effects on radiation dose and image quality using experimental and clinical data, *AJR*, 180, 407–11, 2003.
64. Heaney, D. E., and Norvill, C. A. J., A comparison of reduction in CT dose through the use of gantry angulations or bismuth shields, *Austral. Phys. Eng. Sci. Med.*, 29, 172–78, 2006.
65. Hopper, K. D., et al., Radioprotection to the eye during CT scanning, *Am. J. Neurol. Radiol.*, 22, 1194–98, 2001.
66. International Commission on Radiological Protection, Pregnancy and medical radiation, ICRP Publication 84, *Annals of the ICRP*, 30 (1), 2000.
67. Wagner, L. K., Lester, R. G., and Saldano, L. R., *Exposure of the Pregnant Patient in Diagnostic Radiations: A Guide to Medical Management*, 2nd ed. Madison, WI: Medical Physics Publishing, 1997.
68. Otake, M., Schull, W. J., and Lee, S., Threshold for radiation-related severe mental retardation in prenatally exposed A-bomb survivors: A reanalysis, *Int. J. Radiat. Biol.*, 70, 755–63, 1996.
69. Harrison, R. M., Central axis depth dose data for diagnostic radiology, *Phys. Med. Biol.*, 26, 657–70, 1981.
70. Institute of Physical Scientists in Medicine, *National Protocol for Patient Dose Measurements in Diagnostic Radiology*, prepared by Dosimetry Working Party of IPSM, published by NRPB, Didcot, 1992.
71. Harrison, R. M., Backscatter factors for diagnostic radiology (1–4 mm Al HVL), *Phys. Med. Biol.*, 27, 1465–74, 1982.
72. Klevenhagen, S. C., Experimentally determined backscatter factors for x-rays generated at voltages between 16 and 140 kV, *Phys. Med. Biol.*, 34, 1871–82, 1989.
73. Petoussi-Henss, N., et al., Calculation of backscatter factors for diagnostic radiology using Monte Carlo methods, *Phys. Med. Biol.*, 43, 2237–50, 1998.
74. Tapiovaara, M. J., Lakkisto, M., and Servomaa, A., *PCXMC V1.5: A PC-Based Monte Carlo Program for Calculating Patient Doses in Medical X-ray Examinations*, STUK-A139, Helsinki, Finland: Radiation Nuclear Safety Authority, 2002.
75. International Electrotechnical Commission, *Medical Electrical Equipment: Particular Requirements for Safety-X-ray Equipment for Computed Tomography*, IEC 60601-2-44, Ed. 2.1. Geneva: IEC, 2002.
76. Kalender, W. A., A PC program for estimating organ dose and effective dose values in computed tomography, *Eur. Radiol.*, 9, 555–62, 1999.
77. Stamm, G., and Nagel, H. D., *CT-Expo*, Version 1.5, a tool for dose evaluation in computed tomography, 2005. Available from George Stamm (e-mail: stamm. georg@mh-hannover.de).
78. Shleien, B., Slaback, L. A., and Birky, B. K., *Handbook of Health Physics and Radiological Health*, 3rd ed. Philadelphia: Williams & Wilkins, 1998.
79. Kicken, P. J., and Bos, A. J. J., Effectiveness of lead aprons in vascular radiology: Results of clinical measurements, *Radiology*, 197, 473–78, 1995.

80. Johnson, D. R., et al., Radiation protection in interventional radiology, *Clin. Radiol.*, 56, 99–106, 2001.

81. Vano, E., et al., Lens injuries induced by occupational exposure in non-optimized interventional radiology laboratories, *Br. J. Radiol.*, 71, 728–33, 1998.

82. Vano, E., et al., Radiation exposure to medical staff in interventional and cardiac radiology, *Br. J. Radiol.*, 71, 954–60, 1998.

83. Haskal, Z. J., and Worgul, B. V., Interventional radiology carries occupational risk for cataracts, *RSNA News*, 14, 5–6, 2004.

84. Kato, R. K., et al., Radiation dosimetry at CT fluoroscopy: Physician's hand dose and development of needle holders, *Radiology*, 201, 576–78, 1996.

85. Nawfel, R. D., et al., Patient and personnel exposure during CT fluoroscopy-guided interventional procedures, *Radiology*, 216, 180–84, 2000.

86. International Electrotechnical Commission, *Protective Devices against Diagnostic Medical X-radiation*, Part 2, *Protective Glass Plates*, 61331-2. Geneva: IEC, 1994.

87. International Electrotechnical Commission, *Protective Devices against Diagnostic Medical X-radiation*, Part 3, *Protective Clothing and Protective Devices for Gonads*, IEC 61331-3. Geneva, Switzerland: IEC, 1998.

88. National Council on Radiation Protection and Measurements, *Structural Shielding Design and Evaluation for Medical Use of X-Rays and Gamma Rays of Energies up to 10 MeV*, NCRP Report 49. Bethesda, MD: NCRP, 1976.

89. Simpkin, D. J., and Dixon, R. L., Secondary shielding barriers for diagnostic x-ray facilities: scatter and leakage revisited, *Health Phys.*, 74, 350–65, 1998.

90. British Institute of Radiology and Institute of Physics and Engineering in Medicine, *Radiation Shielding for Diagnostic X-rays*, edited by D.G. Sutton and J.R. Williams. Huddersfield, UK: Charlesworth Group, 2000.

91. National Council on Radiation Protection and Measurements, *Radiation Protection in Dentistry*, NCRP Report 145, Bethesda. MD: NCRP, 2003.

92. National Council on Radiation Protection and Measurements, *Structural Shielding Design for Medical X-Ray Imaging Facilities*, NCRP Report 147. Bethesda, MD: NCRP, 2004.

93. Archer, B. R., Thornby, J. I., and Bushong, S. C., Diagnostic x-ray shielding design based on an empirical model of photon attenuation, *Health Phys.*, 44, 507–17, 1983.

94. Simpkin, D. J., Shielding a spectrum of workloads in diagnostic radiology, *Health Phys.*, 61, 259–61, 1991.

95. Simpkin, D. J., Evaluation of NCRP Report No. 49 assumptions on workloads and use factors in diagnostic radiology facilities, *Med. Phys.*, 23, 577–84, 1996.

96. Simpkin, D. J., Transmission data for shielding diagnostic x-ray facilities, *Health Phys.*, 68, 704–9, 1995.

97. Dixon, R. L., and Simpkin, D. J., Primary shielding barriers for diagnostic x-ray facilities: A new model, *Health Phys.*, 74, 181–89, 1998.

98. Simpkin, D. J., Scatter radiation intensities about mammography units, *Health Phys.*, 70, 238–45, 1996.

99. Sharp, C., Shrimpton, J. A., and Bury, R. F., *Diagnostic Medical Exposures: Advice on Exposure to Ionising Radiation during Pregnancy.* Oxford: National Radiological Protection Board, 1998. Also available online at the NRPB website at www.nrpb.org/publications/misc publications/advice during pregnancy.pdf.

7

Radiation Protection in Nuclear Medicine

Raymond Budd

CONTENTS

This chapter gives an overview of the radiation protection principles applicable to the nuclear medicine department. After a brief introduction into the practice of nuclear medicine and its overall role within medicine, the properties of commonly used radionuclides are discussed. This is followed by an outline of the appropriate methods to limit exposure from radiation hazards and a description of the methods used for the estimation of dose to both patient and staff. Finally, the radiation protection requirements for the facilities provided within the nuclear medicine department and the procedures adopted by staff are discussed. In this regard, the reader is also directed to important publications by international radiation protection bodies.[1,2] Additional radiation protection issues may arise in positron emission tomography (PET) centers with their own cyclotron. These are beyond the scope of the present chapter but have been addressed specifically in several recent articles.[3–5]

7.1   Introduction to Nuclear Medicine

7.1.1   Fundamental Concepts

Nuclear medicine is a branch of medicine that utilizes unsealed radioactive material in a range of complex in vivo procedures for the diagnosis and the treatment of disease.

In diagnosis, it provides functional information as well as structural detail about particular organs and tissues in the body. For this application, a radionuclide is labeled to a specific compound to form a radiopharmaceutical that is designed to target a particular organ or tissue. A prescribed activity is then administered to the patient intravenously, orally, or via inhalation. The radionuclide used in this compound emits gamma ray photons that are of sufficient energy for a large enough number to emerge from the patient to be detected by dedicated imaging equipment such as the gamma camera or single photon emission computed tomography (SPECT) and PET scanners, as depicted in Figure 7.1.

These provide a precise and detailed image of the radioactive distribution throughout the region of interest, and diagnostic information is based

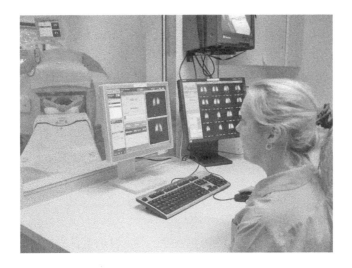

FIGURE 7.1
Typical operation of a SPECT scanner with display of radionuclide images.

upon the accumulation or selective exclusion of the radiopharmaceutical in the particular organ or tissue. Comparison of the concentration and distribution of radiotracer relative to the known normal appearances is the basis for any clinical interpretation. To complement these imaging studies, nuclear medicine procedures can also involve the analysis of biological specimens in the laboratory.

In treatment, nuclear medicine is used to deliver radionuclides or radiopharmaceuticals directly to the organ being treated and as such plays a major role in the therapy of cancer and other diseases. In these cases, most of the radiation dose to the tumor target is delivered by short-range beta particles, with only a small fraction by the accompanying gamma radiation. For both diagnostic and therapy procedures, there is an associated level of risk. This is considerably higher in therapeutic applications because of the large activities routinely used. Nonetheless, all investigations undertaken in the nuclear medicine department require the provision of suitably designed and shielded facilities and the implementation of good work practice in order to minimize this risk to nuclear medicine staff, patients, and members of the general public.

7.1.2 Properties of Currently Used Radionuclides

Nearly three thousand nuclides are known, of which approximately twenty-seven hundred are radioactive. Most of the naturally occurring radionuclides (for example, ^{14}C, ^{40}K, ^{226}Ra, ^{238}U) generally have undesirable properties for use in nuclear medicine, such as very long half-life, particulate emissions, and low specific activity. Consequently, none of these naturally occurring

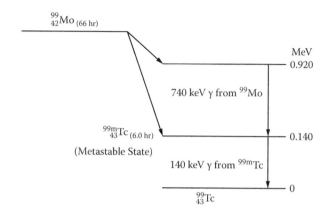

FIGURE 7.2
Simplified decay scheme for ^{99}Mo and ^{99m}Tc.

radionuclides are used for the preparation of radiopharmaceuticals, and all those used in nuclear medicine are artificially produced.

The vast majority of radiopharmaceuticals today utilize ^{99m}Tc as the radionuclide of choice; a simplified scheme of its decay from ^{99}Mo is shown in Figure 7.2. Readers requiring more detailed information for ^{99m}Tc and all other radionuclides are referred to the comprehensive decay schemes published by the Society of Nuclear Medicine.[6]

The widespread use of ^{99m}Tc is a direct consequence of its physical and chemical properties, which make it ideal for organ imaging. These include:

1. Its physical half-life of 6 h, which is long enough to allow imaging up to several hours after its administration, yet short enough to minimize radiation dose to the patient

2. Its decay by isomeric transition, which emits gamma radiation but minimal particulate radiation, also minimizing patient dose

3. Its gamma ray energy of 140 keV, which is high enough to avoid significant attenuation in the body yet is low enough to be easily detected without the need for excessive shielding around the detector

4. Its multiple valence states, which make it suitable for incorporation into a range of radiopharmaceuticals for imaging a range of organs and tissues, such as the skeleton, thyroid, liver, lung, kidney, and so on

5. Its high specific activity, which means that a suitable quantity of ^{99m}Tc can be obtained with only a minimal amount of elemental technetium, minimizing possible toxicity and interference with normal metabolic processes

6. Its relatively inexpensive and convenient production in large quantities for daily use via the ^{99}Mo-^{99m}Tc generator

The longer half-life of 66 h for ^{99}Mo allows its transport over many thousands of kilometers without significant loss of activity. However, its high gamma energy of 740 keV does create a substantial shielding problem during its transport and particularly during its storage in the nuclear medicine laboratory (see Section 7.4.2). In the generator, the molybdenum is bound to an alumina (Al_2O_3) column in the form of $^{99}MoO_4^{2-}$. The technetium activity, in the chemically different form of $^{99m}TcO_4^-$, is not bound by the alumina and is eluted from the column with 5–25 ml of normal saline as Na^+ $^{99m}TcO_4^-$. Typically, 75–85% of the available activity is extracted in a single elution and the maximum activity is available again about 24 h later, although usable quantities can be eluted after about 3–6 h.

Despite the enormous role that ^{99m}Tc plays in the nuclear medicine department, a large number of other radionuclides can be used for special imaging and therapy applications and in clinical research, and the principal characteristics of some of these are listed in Table 7.1.

7.2 Handling of Radioactive Materials

7.2.1 Safety Issues

The safe handling and use of any of these radioactive materials requires an understanding of their associated hazards and the implementation of correct radiation protection principles. In this regard, the hazards from ionizing radiation may be divided conveniently into two classes: those presenting as external hazards and those as internal hazards. External hazards arise from sources outside the body that can irradiate all or part of the body with sufficient energy to affect the skin or underlying tissues. Internal hazards, on the other hand, arise when radioactive materials enter the body through inhalation, injection, ingestion, or absorption through the skin or wounds. The radioactive substance may be rapidly eliminated from the body, or it may selectively concentrate in a particular organ with a resulting absorbed dose. There is always the possibility of contamination and unwanted radiation dose from radioactive material, and it is therefore necessary for all nuclear medicine staff to adopt and employ some practical methods to minimize this exposure.

7.2.2 Methods for Limiting Exposure to External Hazards

The most effective method to minimize exposure to external radiation hazards involves the correct use of time, distance, and shielding as previously discussed in Chapter 6. This is achieved by minimizing the time spent in

TABLE 7.1

Physical Characteristics of Radionuclides Commonly Used in Nuclear Medicine

Nuclide	Half-Life	Approximate Energy, MeV (% abundance)
Gamma Emitters Used for Imaging		
$^{67}_{31}$Ga	78 h	0.093 (40), 0.184 (20), 0.300 (17), 0.393 (5)
$^{99m}_{43}$Tc	6 h	0.140 (90)
$^{111}_{49}$In	2.8 days	0.171 (90), 0.245 (94)
$^{131}_{53}$I	8.0 days	0.284 (6), 0.364 (81), 0.637 (7)
$^{133}_{54}$Xe	5.3 days	0.081 (37)
$^{201}_{81}$Tl	73 h	0.167 (9), x-rays 0.069–0.083 (93)
Positron Emitters Utilizing the 0.511 MeV Photons for Imaging		
$^{11}_{6}$C	20.4 min	0.960
$^{13}_{7}$N	10 min	1.190
$^{15}_{8}$O	2 min	1.730
$^{18}_{9}$F	110 min	0.635
$^{82}_{32}$Rb	1.3 min	3.150
Beta Emitters Used for Therapy		
$^{32}_{15}$P	4.3 days	1.71 max, 0.7 mean beta
$^{89}_{38}$Sr	50.5 days	1.46 max, 0.58 mean beta
$^{90}_{39}$Y	64 h	2.24 max, 0.0.93 mean beta
$^{131}_{53}$I	8.0 days	0.19 mean beta, 0.364 gamma (81)
$^{153}_{62}$Sm	46 h	0.81 max, 0.23 mean beta, 0.103 gamma (96)

proximity to radioactive sources, maximizing the distance between the source and exposed person, and using suitable shielding. Shielding always improves the effectiveness of radiation protection practices and is mandatory in many situations.

Minimizing time: Radiation dose is directly related to the duration of exposure, so if the time of exposure is doubled, the radiation dose will be doubled. Consequently, those involved in diagnostic or therapeutic procedures using radioactive material can keep their radiation dose to a minimum by reducing any unnecessary time spent in close proximity to the source, which in most cases is the patient.

Maximizing distance: As the distance between the source of radiation and a person increases, the intensity of the radiation, and hence the exposure, decreases rapidly. For a point source of radiation, the reduction of exposure obeys the inverse square law. Mathematically, this is expressed as

$$\frac{I_1}{I_2} = \left(\frac{d_2}{d_1}\right)^2$$

where, I_1 is the intensity at distance d_1 and I_2 is the intensity at distance d_2. The beam from the target of an x-ray tube is an example of a point source that accurately obeys this rule. Within the nuclear medicine department, the radiation emerging from a small glass vial containing radioactive material also obeys the inverse square law to a good approximation. However, the dose rate from the distributed source within a nuclear medicine patient does not vary exactly as the inverse square of the distance, but the practice of maximizing distance is still an appropriate way for a nuclear medicine technologist to minimize radiation dose in this case.

Using appropriate shielding: When shielding material is placed between the radiation source and persons exposed, the level of exposure is greatly reduced. In the nuclear medicine department, considerable use is made of lead, and to some extent bricks and reinforced concrete, for this purpose; this will be discussed later (see Section 7.4).

7.2.3 Methods for Limiting Exposure to Internal Hazards

The possibility of contamination and unwanted radiation dose from internal hazards can be minimized by using the practical steps discussed below. Many if not all of these steps are included in local mandatory regulations and standards applicable to the use of unsealed radioactive substances in areas such as the radiopharmaceutical laboratory in the nuclear medicine department.

1. Appropriate protective clothing and eyewear plus suitable rubber or plastic gloves should be worn at all times when preparing radiopharmaceuticals for patient administration (see Figure 7.3).

2. Secondary containment such as trays should be used for all procedures to minimize the possible spread of contamination.

3. A fume cupboard should be available for operations that produce vapor, spray, dust, or radioactive gas.

4. All work with unsealed radioactive substances should be segregated from other laboratory work.

5. New procedures and techniques should be practiced with nonradioactive substances.

6. Working procedures and emergency contingency plans should be regularly reviewed.

7. A high standard of cleanliness should be maintained at all times, and the laboratory benches and floor should be regularly monitored with a radiation detector. The purpose of regular radiation monitoring is to identify areas where significant dose rates may exist or where persons may be contaminated both internally and externally through the spread of unsealed radioactive materials. Using a suitable ionization detector, surface contamination can be located by scanning slowly and systematically over the whole surface, overlapping each scan movement.

FIGURE 7.3
Examples of appropriate protective clothing and safety facilities in the radionuclide laboratory, such as labeled fume cupboard, lead glass barrier, and secondary containment trays.

7.3 Radiation Dosimetry

The administration of a radiopharmaceutical necessarily exposes the patient to a radiation dose. For a diagnostic procedure, the absorbed dose is kept to an acceptably low level consistent with the desired investigative objective in order to minimize any possible detrimental effects. For a therapeutic procedure, on the other hand, the radiation dose needs to be high enough to produce the desired clinical response.

In both circumstances, the patient becomes a mobile source of radiation, which presents a risk to hospital staff, other patients, and family members or other persons with whom the patient has contact following discharge from the hospital. Radiation dosimetry considerations relevant to both of the above perspectives represent an important subject and are discussed in the following sections.

7.3.1 Calculation of Radiation Dose to Staff

Personnel in nuclear medicine departments may receive radiation doses that approach or exceed the recommended limits unless proper precautions are taken. As discussed earlier (see Section 7.2.2), in addition to sensibly minimizing the time spent near a radioactive source, protection against gamma rays may be achieved in two ways: by shielding, usually with lead, and by the use of distance.

The effective dose E received by exposure to an external radioactive source can be determined from the following equation:

$$E = \Gamma A_o \, t \, (1/d^2) \text{ mSv} \tag{7.1}$$

where Γ is the gamma ray dose constant in mSv/MBq-h at 1 m, A_o is the radionuclide activity in megabecquerals, t is the time in hours, and d is the distance in meters.

Table 7.2 provides data for a number of radionuclides adapted from Shleien et al.[7] and includes the Γ value as well as the half-value layer (HVL) and the tenth-value layer (TVL) for lead. The data are used in the example calculations that follow.

Example 7.1

Determine the effective dose received by a nuclear medicine technologist who is exposed to a typical unshielded diagnostic source of 800 MBq of ^{99m}Tc at a distance of 30 cm for a period of 5 min.

From Table 7.2, the Γ for ^{99m}Tc is 3.317×10^{-5} mSv/MBq-h at 1 m. Therefore, the effective dose is

TABLE 7.2

Dose Rate Constants HVL and TVL for Commonly Used Radionuclides

Radionuclide	Γ mSv/MBq-h at 1 m	HVL Lead (cm)	TVL Lead (cm)
^{18}F	1.879×10^{-4}	0.40	1.30
^{67}Ga	3.004×10^{-5}	0.15	0.70
^{99}Mo	3.052×10^{-5}	0.60	20.0
^{99m}Tc	3.317×10^{-5}	0.03	0.09
^{111}In	1.356×10^{-4}	0.10	0.30
^{131}I	7.647×10^{-5}	0.30	1.70
^{133}Xe	2.783×10^{-5}	0.03	0.09
^{201}Tl	2.372×10^{-5}	0.02	0.08

$$E = \Gamma A_o \, t \, (1/d^2)$$

$$= 3.317 \times 10^{-5} \times 800 \times 5/60 \times (1/0.3)^2$$

$$= 0.0246 \text{ mSv i.e., approximately 25 } \mu Sv$$

Example 7.2

Consider a scenario similar to the previous example except that the unshielded source is now a typical therapeutic source of 3,700 MBq of ^{131}I.

The Γ for ^{131}I is 7.467×10^{-5} mSv/MBq-h at 1 m. Therefore, the effective dose is

$$E = \Gamma A_o \, t \, (1/d^2)$$

$$= 7.647 \times 10^{-5} \times 3700 \times 5/60 \times (1/0.3)^2$$

$$= 0.262 \text{ mSv i.e., approximately 260 } \mu Sv$$

These examples clearly demonstrate that significant occupational exposure can readily occur from unshielded radioactive sources, particularly in therapeutic applications because of the larger activities used.

It is often assumed that the larger occupational effective dose received by nuclear medicine technologists compared to other health professionals, typically of the order of 3 mSv/year compared to <1 mSv/year for radiographers, for example, is the result of their routine involvement with the preparation of radiopharmaceuticals in the laboratory rather than with patient-oriented tasks. A number of investigations, however, have confirmed that the primary source of exposure is in fact from the radioactive patient.[8,9] In particular, Smart[10] has specifically identified the extra time associated with transferring incapacitated patients, assisting difficult injection procedures, and setting up myocardial scans as the major contributors to occupational exposure for nuclear medicine personnel.

However, activities undertaken in the nuclear medicine laboratory still account for significant occupational exposure, and the provision of a safe environment is essential in order to minimize this exposure (see Section 7.4). Obviously, for the handling and storage of gamma-emitting radionuclides, some form of shielding is normally required, and lead is the most common choice because of its high density and high atomic number. The transmission of gamma rays through lead will vary with their energy and thus will be different for each radionuclide. The gamma ray absorption properties of lead are most conveniently expressed in terms of the HVL and the TVL. It can seen in Table 7.2 that for radionuclides that emit essentially only one energy gamma ray, and therefore have an approximately exponential attenuation in material, such as ^{99m}Tc and Xe-133, the TVL is 3.3 times the HVL. By contrast, for a radionuclide that emits a number of gamma rays of different energies, such as Ga-67 and I-131, the TVL will scale more dramatically, since at high filtration the penetration is determined primarily by the higher-energy gamma rays. The appropriate thickness of shielding for a given activity of a particular radionuclide can be determined using the values of HVL and TVL provided in Table 7.2.

7.3.2 Calculation of Radiation Dose to the Patient

In nuclear medicine investigations, it is impractical if not impossible to measure the radiation dose to specific organs directly using any kind of radiation detector. Patient dose has to be estimated by first establishing the relevant biological uptake, distribution, and excretion data, which are often extrapolated from animal or limited human data, then using the known physical data for the specific radionuclide and applying this information to specially developed dosimetric equations.

The mean absorbed dose $D(t \leftarrow s)$ expressed in grays to a target organ (t) from a radionuclide distributed uniformly in a source organ (s) has been formulated by the Medical Internal Radiation Dose (MIRD) Committee[11] as

$$D(t \leftarrow s) = (\tilde{A}_S/m) \, \Sigma_i \, [\Delta_i \, \varnothing_i(t \leftarrow s)] \qquad (7.2)$$

where $\tilde{A}_S$ is the cumulated activity (i.e., the total number of transformations) in source organ (s), m is the mass of the target organ (t), Δ_i is the equilibrium dose constant for particles or photons of a particular type and energy, here indicated by i, and $\varnothing_i(t \leftarrow s)$ is the absorbed fraction and represents the fraction of the energy of type i emitted by the source organ that is absorbed in the target organ.

In the majority of problems encountered in nuclear medicine, the radio-activity is distributed in the target organ itself. In these cases, the absorbed fraction is self-absorbed and is expressed simply as $\emptyset_i$. Determination of the absorbed fraction requires knowledge of the interaction of radiation with matter. In the case of particulate radiations almost all of the energy emitted by a radionuclide is absorbed in the source organ, provided the source volume is larger than 1 cm^3. In this case, $\emptyset_i(t \leftarrow s) = 0$, unless (t) and (s) are the same, in which case $\emptyset_i = 1$. This also holds true for x and γ radiations of energy less than 11 keV.

Equation (7.2) can be rewritten as

$$D(t \leftarrow s) = \tilde{A}_s \, \Sigma_i [\Delta_i \, \Phi_i(t \leftarrow s)] \tag{7.3}$$

where $\Phi_i(t \leftarrow s)$ is the specific absorbed fraction that depends on the radiation type and the size, shape, and separation of the source and target organs, and is defined by $\Phi_i(t \leftarrow s) = \emptyset_i(t \leftarrow s)/m$.

The MIRD dosimetry scheme utilizes the S factor, defined as $S(t \leftarrow s) = \Sigma_i [\Delta_i \, \Phi_i(t \leftarrow s)]$, and values for S have been calculated and published for radionuclides and source–target configurations commonly used in nuclear medicine.[12] The mean absorbed dose from a specified radionuclide can therefore be calculated by the simplified equation

$$D(t \leftarrow s) = \tilde{A}_s \, S(t \leftarrow s) \tag{7.4}$$

Since there are generally a number of source organs, the total mean absorbed dose to target organ (t) is given by

$$D(t) = \Sigma \, D(t \leftarrow s) = \Sigma \, \tilde{A}_s \, S(t \leftarrow s) \tag{7.5}$$

In order to use the published tables to compute the absorbed dose, it is necessary to calculate the cumulated source activity in the organ(s), which is

$$\tilde{A}_s = \int_0^\infty A(t)dt$$

over the time interval of interest. This simplifies to $\tilde{A}_s = A_0 (1.443 \, T_{eff})$, where A_0 is the initial activity administered at time $t = 0$ and T_{eff} is the effective half-life, which is approximately related to the physical half-life of the radionuclide T_p and the biological half-life of the radiopharmaceutical T_b by

$$\frac{1}{T_{eff}} = \frac{1}{T_p} + \frac{1}{T_b}$$

TABLE 7.3

Values of S for ^{99m}Tc (mGy/MBq-s)

Target Organs	Source Organs		
	Liver	Spleen	Bone Marrow
Liver	3.23×10^{-6}	7.20×10^{-8}	8.93×10^{-8}
Spleen	7.20×10^{-8}	2.33×10^{-5}	9.17×10^{-8}
Bone marrow	8.29×10^{-8}	8.41×10^{-8}	1.74×10^{-6}

The effective half-life is always less than the shorter of T_p or T_b, and if $T_p \gg T_b$, then $T_{eff} \approx T_b$, and if $T_b \gg T_p$, then $T_{eff} \approx T_p$.

Example 7.3

Suppose that 200 MBq of ^{99m}Tc-sulfur colloid is administered to a patient for bone marrow imaging and is rapidly distributed to the liver (70%), spleen (10%), and bone marrow (20%). Calculate the absorbed dose to the liver, spleen, and bone marrow using the S values listed in Table 7.3.

In order to use the general equation for estimating absorbed dose, $D(t) = \Sigma \tilde{A}_S S(t \leftarrow s)$, it is necessary to calculate the cumulated activities for each source organ. Since the biological half-life of the colloid can be assumed to be long, the effective half-life is therefore equal to the physical half-life of 6 h (21,600 s) and the initial activity A_0 is equal to 200 MBq.

Therefore,

$$\tilde{A}_{liv} = 0.70 \times 200 \times 21600 \times 1.443 = 4.36 \times 10^6 \text{ MBq-s}$$

$$\tilde{A}_{spl} = 0.10 \times 200 \times 21600 \times 1.443 = 6.23 \times 10^5 \text{ MBq-s}$$

$$\tilde{A}_{bm} = 0.20 \times 200 \times 21600 \times 1.443 = 1.25 \times 10^6 \text{ MBq-s}$$

The estimated dose to the liver is

$$D(liv) = \tilde{A}_{liv} \, S(liv \leftarrow liv) + \tilde{A}_{spl} \, S(liv \leftarrow spl) + \tilde{A}_{bm} \, S(liv \leftarrow bm)$$

$$= (4.36 \times 10^6) \times (3.23 \times 10^{-6}) + (6.23 \times 10^5) \times (7.20 \times 10^{-8})$$

$$+ (1.25 \times 10^6) \times (8.93 \times 10^{-8})$$

$$= 14.08 + 0.045 + 0.112 = 14.24 \text{ mGy}$$

The dose to the spleen is

$$D(spl) = \tilde{A}_{liv}\, S(spl \leftarrow liv) + \tilde{A}_{spl}\, S(spl \leftarrow spl) + \tilde{A}_{bm}\, S(spl \leftarrow bm)$$

$$= (4.36 \times 10^6) \times (7.20 \times 10^{-8}) + (6.23 \times 10^5) \times (2.33 \times 10^{-5}) +$$

$$(1.25 \times 10^6) \times (9.17 \times 10^{-8})$$

$$= 0.314 + 14.52 + 0.115 = 14.95 \text{ mGy}$$

The dose to the bone marrow is

$$D(bm) = \tilde{A}_{liv}\, S(bm \leftarrow liv) + \tilde{A}_{spl}\, S(bm \leftarrow spl) + \tilde{A}_{bm}\, S(bm \leftarrow bm)$$

$$= (4.36 \times 10^6) \times (8.29 \times 10^{-8}) + (6.23 \times 10^5) \times (8.41 \times 10^{-8}) +$$

$$(1.25 \times 10^6) \times (1.74 \times 10^{-6})$$

$$= 0.361 + 0.052 + 2.18 = 2.59 \text{ mGy}$$

Similarly, the absorbed dose received by other organs can be estimated using the appropriate S factors and the combined information used to determine the effective dose to the patient from this investigation. To do this, the equivalent dose H_T to each organ is obtained by multiplying the absorbed dose by the radiation weighting factor w_R, which is defined as unity for radiations from ^{99m}Tc and for all other radiations utilized in nuclear medicine. The individual equivalent doses are then multiplied by the appropriate tissue weighting factor w_T for each organ[13] to give values of $H_T w_T$, which are finally summed to determine the effective dose. These results are presented in Table 7.4.

The effective dose from this investigation is therefore approximately 1.9 mSv. A modification to tissue weighting factors is one of several changes currently proposed by the ICRP in its latest draft recommendations,[14] and any calculation of effective dose will need to incorporate these factors once they have been published and adopted by local regulations. The calculation method described above can be used for all diagnostic and therapy investigations using the relevant published S values.

7.3.3 Radionuclide Diagnostic Reference Activities

The principle of optimization implies that all medical radiation doses should be as low as reasonably achievable. In radiology, this can be achieved by setting diagnostic reference levels (DRLs) for specific procedures, and DRLs have been determined and documented for specific radiological investigations in some jurisdictions to provide the optimum diagnostic information, that is, the maximum benefit to the patient for the least risk (see Chapter 6).

In nuclear medicine, similar reference levels have been introduced in a number of countries with the aim of optimizing nuclear medicine proce-

TABLE 7.4

Estimation of Effective Dose

Target Organ	Equivalent Dose, H_T mSv	Tissue Weighting Factor, w_T	HTwT mSv
Liver	14.24	0.05	0.712
Spleen	14.95	0.05	0.748
Bone marrow	2.59	0.12	0.311
Gonads	0.42	0.20	0.084
Colon	.0.36	0.12	0.043
Lung	1.04	0.12	0.125
Stomach	1.20	0.12	0.144
Bladder	0.20	0.05	0.010
Breast	0.50	0.05	0.025
Esophagus	0.50	0.05	0.025
Thyroid	0.15	0.05	0.038
Bone surfaces	1.25	0.01	0.013
Remainder	1.00	0.01	0.010
		Total	1.92

dures. The Administration of Radioactive Substances Advisory Committee (ARSAC) has published guidelines for good clinical practice for nuclear medicine in the United Kingdom.[15] These guidelines include tabulated DRLs in the form of recommended administered activities for all current radionuclide investigations. Similarly, the Australian and New Zealand Society of Nuclear Medicine (ANZSNM) and the Australasian Radiation Protection Society (ARPS) have jointly developed diagnostic reference levels for adults[16] and children.[17]

7.3.4 Implications of Pregnancy for Patients and Staff

7.3.4.1 Pregnant Patients

The effects of radionuclides on the developing embryo or fetus have not been studied as extensively as the consequences of externally administered x-rays. This is unfortunate, since the situation is much more complex in nuclear medicine investigations on pregnant women where the fetal irradiation will result from radioactivity that is located in nearby maternal organs and, in some investigations, from radioactivity that has actually transferred across the placenta. This is of particular concern in therapy procedures involving ^{131}I because of uptake in the fetal thyroid. Consequently, the possibility of pregnancy should be considered for all women of childbearing age. Multilingual posters and notices should be clearly displayed in the nuclear medicine department instructing patients to inform staff members if there is the possibility that they may be pregnant. An informed decision can then be made as to whether the procedure is still advisable or whether it can be replaced with an alternative nonionizing procedure or delayed to a more appropriate time.

In cases of inadvertent fetal irradiation or when a pregnant patient needs an investigation involving ionizing radiation, estimates of fetal dose from an extensive range of radiopharmaceuticals can be obtained from Wagner et al.[18] This reference also includes management guidelines and recommendations applicable to the pregnant patient in such cases, as does the International Commission on Radiological Protection (ICRP) Publication 84.[19] Absorbed dose estimates from specific radiopharmaceuticals are also available from the ICRP for the evaluation of the risks versus benefits, and to provide adequate information to the patient and the referring physician.[20]

Alternatively, the availability of the pregnant female phantom series and its incorporation into dosimetry software models such as MIRDOSE 3 and more recently OLINDA has made possible the estimation of absorbed doses to the fetus from radionuclides in the body at different stages of gestation.[21–23] These can provide an informed evaluation of the associated risks and benefits of the different procedures.

When a radiopharmaceutical is administered to a breast-feeding patient, it may be secreted in her milk with the nursing child receiving a radiation dose from the ingested radioactivity. The secreted radioactivity will decrease rapidly with time after administration; however, it is recommended that breast feeding be discontinued for a limited period after any nuclear medicine procedure and in some cases be ceased altogether.[24]

7.3.4.2 Pregnant Staff

A female member of staff in a nuclear medicine department should, on becoming aware that she is pregnant, notify her employer. Working conditions should then be reviewed and adapted if necessary to ensure that the level of radiation protection given to the embryo is the same as that required for the general public. The 1990 and 2007 recommendations of the ICRP are that a fetus should not receive more than 1 mSv during the declared term of the pregnancy. This can be interpreted as broadly equivalent to a dose at the surface of the abdomen of a pregnant woman of about 2 mSv for x-rays, but a lower level, possibly 1.3 mSv, for higher-energy radiation from radionuclides such as ^{99m}Tc and ^{131}I.[25] Even if the woman's individual dose is estimated to be below 1.3 mSv, it would be advisable for a pregnant nuclear medicine technologist to avoid preparing and administering therapy doses and imaging very ill patients.

7.3.5   Maladministration of a Radiopharmaceutical

Even with well-established patient management procedures in place, there is still the potential for an incorrect administration of radioactivity because of human error. This occurs when a patient receives the wrong radiopharmaceutical or the incorrect activity for the requested procedure, or when the radioactivity is administered to the wrong patient. Mandatory confirmation and cross-checking of all details for a requested procedure should ensure that mistakes are rare. However, if and when they do occur, the priority is to implement any measures that will enhance the excretion of the radiopharmaceutical so as to minimize the resulting dose. It is obligatory that the patient and the referring physician are immediately informed that an incorrect administration has occurred and that the hospital and regulatory authorities are notified.

7.4   Overview of Department Design Requirements

7.4.1   Laboratory Protection Equipment

In order to minimize the external exposure of the nuclear medicine technologist, the radiopharmaceutical laboratory should be equipped with at least the following items, many of which are depicted in Figure 7.3:

- Clear and appropriate use of the radiation trefoil sign
- Fume cupboard
- Protective clothing and eyewear
- Rubber or plastic gloves
- Lead barrier and lead glass window at the radionuclide draw-up station
- Secondary containment trays
- Tongs and forceps for remote handling of radionuclides
- Lead containers and pots for radionuclide storage
- Syringe shields
- Secure and shielded storage cabinets
- Radiation detector

7.4.2   Imaging and Laboratory Rooms

Shielding of rooms in the nuclear medicine department is designed to ensure that dose constraints for staff and members of the public are not exceeded, and also to prevent degradation of nuclear medicine images by radiation emitted from patients in adjoining locations, such as the injection room or the waiting room. Since the majority of planar and SPECT nuclear medicine imaging utilizes ^{99m}Tc, a lead thickness of 2 mm is routinely recommended for shielding in the walls of all rooms, including imaging rooms and the laboratory. This is equivalent to two TVLs (see Table 7.2), providing an attenuation of approximately 99%. If the imaging room has a viewing window, as shown in Figure 7.1, it is recommended that the glass also have a lead equivalence of 2 mm.

^{99m}Tc is routinely available from a transportable ^{99}Mo-^{99m}Tc generator and the parent radionuclide, ^{99}Mo, requires more stringent radiation protection because of its relatively long half-life of 66 h and particularly because of its high-energy gamma photons of 740 keV, with a 20 mm TVL for lead. Consequently, the generator needs to be housed in a specially designed facility with total shielding up to 80 mm lead for activities of the order of tens of gigabecquerals.

7.4.3   Radionuclide Therapy Rooms

Therapy rooms are often custom designed with walls constructed of brickwork or solid concrete to provide adequate shielding for the high activities of ^{131}I usually employed. Floors should be covered with continuous vinyl sheets that are coved at the walls. Special facilities that should also be provided include a hand-wash basin at the entrance for staff and a separate hand basin, shower, and toilet for the patient. All pipes leading from the toilet should be shielded and routed either directly into the main sewer or to shielded storage tanks. The actual format depends on the total water outflow from the hospital and local regulations regarding the level of radioactivity that can be released directly into the sewer. If storage tanks are to be used, the residual ^{131}I waste can be disposed of following the principle of delay and decay, discussed in Section 7.5.4.3. In practice, this will require several tanks to which influx can be switched.

7.4.4   PET and PET-CT Facilities

The increasing use of PET and the recent introduction of PET-CT scanners provide special considerations with regard to facility shielding. The factors that affect the style and level of this shielding include the location of the PET center, for instance, whether it is integrated within the nuclear medicine department or whether it is in a separate self-contained site. If it is the former, attention needs to be given to the additional shielding requirements for rooms containing conventional gamma cameras and SPECT scanners so that their function is not compromised by the presence in the department of patients containing fluorine-18. Other important issues relevant to both locations are the number of patients scanned, the activity of ^{18}F used, and the time each patient spends in the department. Shielding requirements for CT scanners have been covered in Chapter 6 but in the case of PET/CT systems, consideration need only be given to the PET component because the shielding implications of the 0.511 MeV annihilation photons from ^{18}F and other positron-emitting radionuclides are far greater than that for the x-rays produced by the CT scanner.

The TVL of 0.511 MeV photons in lead is approximately 13 mm. The sole use of this material for shielding would be prohibitively expensive, and so shielding is usually provided with concrete (TVL ~ 19 cm), sometimes in combination with sheets of lead and even iron (TVL ~ 6 cm). The discussion of specific examples of shielding scenarios is beyond the scope of his book, but the interested reader is directed to an excellent and comprehensive coverage of this topic published by the American Association of Physicists in Medicine (AAPM).[26]

7.5   Environmental Safety Considerations

7.5.1   Special Considerations Governing the Discharge of Patients

In general, a patient administered no more than the ARSAC diagnostic reference-level activity need not adhere to any restrictions with regards to contact with family members or with the general public. For therapeutic applications, however, this will not necessarily apply, as there is potentially a greater risk involved because of the higher activities present. In the case of ^{131}I therapy, for example, this will include the possible contamination from radioactive tissue as well as irradiation from the patient's emitted radiation. Radioactive iodine in sweat, saliva, and urine presents a potential risk of contamination, but normal hygiene precautions will in general provide adequate protection from these hazards. Protection from the emitted radiation is usually achieved by treating the patient in the hospital in a specially designed treatment room (see Section 7.4.3). The length of stay is typically at least 3 days, and monitoring of the patient prior to discharge is essential.

The ICRP has published recommendations for the release of patients following therapy with unsealed radionuclides.[27] These are designed so that the radiation dose to persons with whom the patient may make contact outside the hospital (i.e., members of the public, family members, or carers) is kept as low as reasonably achievable, taking into account the particular social and economic factors, and also that it does not exceed the relevant dose limit prescribed by local regulations.

7.5.2   Disposal of a Radioactive Corpse

If a patient dies after the administration of radioactivity for a diagnostic procedure, usually no special handling precautions are required in addition to those routinely adopted by those involved in postmortem examinations or embalming procedures. The only exception is if death occurs within the first day of administration, when an assessment of the radiation hazard would be required. However, the situation is quite different if the patient had recently received a therapeutic level of activity, particularly ^{131}I; the typical precautions required in this situation have been reported elsewhere.[28,29] In some cases it may be necessary to store the deceased patient for a limited period to comply with local regulations.

7.5.3 Transport and Storage of Radioactive Material

In most countries the transportation of radioactive materials is subject to strict legal control, and regulations are largely derived from the published recommendations of the International Atomic Energy Agency (IAEA).[30] The principal aim of these regulations is to "establish standards of safety which provide an acceptable level of control of the radioactive hazards to persons, property and the environment that are associated with the transport of radioactive material."

The transport and delivery of radioactive materials from their production sites to the nuclear medicine department is generally the responsibility of the radionuclide supplier. However, once the material is on site, its storage and internal movement become the responsibility of the hospital or department. All radioactive material should be securely shielded and packaged with a clear description of the contents attached. When radioactive materials are appropriately and safely contained within a secure and shielded container, they represent minimal hazard to members of staff, patients, and the general public. However, the method of transportation should always take account of any foreseeable accident, being mindful that within the hospital, accidental dropping of a patient radiopharmaceutical preparation may occur as it is moved from the laboratory to the injection room, or throughout the hospital in cases where patients are injected outside the nuclear medicine department, such as in the ward.

7.5.4 Disposal of Radioactive Waste

Radioactive waste can be managed in a manner acceptable to society, and the objectives of waste management practices are to ensure that wastes do not pose environmental problems in either the short or long term, and that no one is subject to any significant health risk. Effective waste management is based on three principles:

1. Delay and decay
2. Dilute and disperse
3. Concentrate and contain

7.5.4.1 Delay and Decay

Most radioactive waste produced in a medical facility is in very small quantities of short-lived radionuclides. In such cases, the wastes can be stored until the activity decays to such a level that they can be considered nonradioactive and dumped with approval of the local licensing authority along with inactive waste. This technique may be used for solids, liquids, and gases, and metal drums should be used for this dispersal method. All containers must have a label on the lid that states the starting time, the date, and estimated activity of each deposit, the date the lid was sealed, and the name and signature of the designated responsible person.

7.5.4.2 Dilute and Disperse

This is the deliberate release of liquid waste into the environment with dilution by water to sufficiently low levels. For low-level liquid waste generated in nuclear medicine departments, the most convenient and widely used practice for disposal is to discharge the wastes into sewers under controlled conditions. This can only be done at specially designated sinks, which should be clearly and permanently labeled for radiation use. The specific activity (megabecquerals per cubic meter) must not exceed the limits specified by the local regulatory authority. The waste should be diluted into a large container before slowly pouring it into the sink, being careful to avoid splashing. The tap should be left running for a few minutes to fully flush the waste out of the immediate laboratory drain pipes. Because of its volatility, I-125 that has been diluted into large liquid volumes should be disposed of via a sink in a fume cupboard, being careful to avoid creating a vapor. If it has been dispensed into small sealed tubes, it may be disposed of along with other solid waste.

7.5.4.3 Concentrate and Contain

This practice can be applied to liquid wastes through chemical treatment processes and solid waste by volume reduction techniques, the aim being to minimize volumes and isolate and store the active component.

7.5.5 Accident Contingency Plans

In the case of a radiation emergency in the nuclear medicine department, the nature and extent of any potential or actual risk to personnel will vary considerably. Minor spillage of radioactivity may occur, but the frequency and consequences can be minimized by good laboratory practice, particularly the use of secondary containment as indicated in Section 7.2.2. A more common event is radioactive contamination resulting from an incontinent patient. Here the correct and prompt response to such accidents should ensure that the contamination does not spread beyond the immediate area of the spill.

The ready availability of a spill-pack containing plastic gloves, disposable overshoes, absorbent material, and a large plastic bag for the collection of contaminated articles will greatly aid this process. In the unlikely but still possible occurrence of a more serious accident involving personnel injury, the general rule for the priority of action is: *treat, delineate, contain, decontaminate, and report.* In all situations, though, any personnel decontamination should proceed a cleanup of the site.

7.5.5.1 Treat

If anyone has been injured in the accident, then the provision of treatment is the first priority. The person providing this should immediately call for assistance and then, wearing rubber or plastic gloves, treat any injuries taking care to minimize the spread of contamination. In the extreme situation of a life-threatening injury, medical treatment takes priority over treatment for contamination. All contaminated skin surfaces should be cleaned and contaminated clothing removed as cuts or damage to skin allow the opportunity for radioactive material to be internalized. Special care must be used when cleaning skin surfaces as there is a risk of contamination entering the bloodstream through cuts or abrasions. Liberal amounts of wet nonabrasive soap should be applied while keeping to the contaminated area only. After rinsing with warm water and drying, the relevant area should be remonitored. If contamination is still present, the area should be washed again with a soft bristle brush, being careful not to damage the skin, followed by rinsing, drying, and remonitoring. If the hands are widely contaminated, particular attention should be given to palms, creases, around and under fingernails, and between fingers. If the skin is still contaminated, a stronger detergent may be used, but damage to the skin surface must be avoided at all times.

7.5.5.2 Delineate

The contaminated area should be clearly delineated, noting that this is generally much larger than first estimated. The extent of the spill can be assessed with the use of an appropriate radiation contamination monitor.

7.5.5.3 Contain

The spill should be covered with absorbent plastic-backed pads or paper towels and access restricted to the contaminated area. Depending on the extent of the spill, it may be appropriate to set up a decontamination zone with plastic sheets for people involved in the decontamination so they may change overshoes and gloves before entering or leaving the area.

7.5.5.4 Decontaminate

The liquid spill should then be cleaned up with absorbent material. When no visible spilled material remains, the affected area should be monitored to check the progress of the decontamination. The decontamination process should be continued as necessary until the contamination is reduced to the minimum level. All materials used in the decontamination process should be treated as radioactive waste and placed in a separate plastic bag. Finally, all persons involved in the spillage or decontamination operation should be monitored.

7.5.5.5 Report

A written report of the accident should be sent to the appropriate department or hospital person, usually the designated radiation protection officer, who will decide whether it warrants further reporting depending on local statutory regulations.

7.5.6 Mandatory Record Keeping

The keeping of complete and up-to-date records is a statutory requirement under many licensing arrangements for the use of radioactive materials. The records to be kept include:

- Inventory of unsealed sources with details of radionuclide, activity, chemical form, date of receipt, and place of storage
- Record of radionuclide use with details of the stock solution, activity, volume, purpose, time, and date
- Record of radionuclide disposal with details of method (e.g., flushing into sewer system or disposal as dry or liquid waste), estimated activity (obviously it will be impossible to record accurate activities), and date
- Records of area surveys with details of the date, area surveyed, purpose of survey (e.g., surface contamination), radiation detector used, and results

References

1. International Atomic Energy Agency, *Applying Radiation Safety Standards in Nuclear Medicine*, Safety Reports Series 40. Vienna, Austria: IAEA, 2005.
2. International Commission on Radiological Protection, Radiological protection and safety in medicine, ICRP Publication 73, *Annals of the ICRP*, 26, 2, 1996.
3. Hertel, N. E., Shannon, M. P., Wang, Z. L., Valenzano, M. P., Mengesha, W., and Crowe, R., Neutron measurements in the vicinity of a self-shielded PET cyclotron, *Radiation Protection Dosimetry*, 108, 255–61, 2004.
4. Pevey, R., Miller, L. F., Marshall, B. J., Townsend, L. W., and Alvord, B., Shielding for a cyclotron used for medical isotope production in China, *Radiation Protection Dosimetry*, 115, 415–19, 2005.
5. Barquero, R., Méndez, R., Martí-Climent, J. M., and Quincoces, G., Monte Carlo neutron dose estimations inside a PET cyclotron vault room, *Radiation Protection Dosimetry*, in press, 2007. Advance access published online on May 15, 2007 doi:10.1093/rpd/ncm096 available at http://rpd.oxfordjournals.org/cgi/content/abstract/ncm096v1.
6. Weber, D. A., Eckerman, K. F., Dillman, L. T., and Ryman, J. C., *MIRD: Radionuclide Data and Decay Schemes*. New York: Society of Nuclear Medicine, 1989.

7. Shleien, B., Slaback, L., and Birky, B., *Handbook of Health Physics and Radiological Health*, 3rd ed. Philadelphia: Williams & Williams, 1998.
8. Benata, N. A., Cronin, B. F., and O'Doherty, M. J., Radiation dose rate from patients undergoing PET: Implications for technologists and waiting areas, *European Journal of Nuclear Medicine*, 27, 583–89, 2000.
9. McElroy, N. L., Worker dose rate analysis based on real time dosimetry, *Health Physics*, 74, 608–9, 1998.
10. Smart, R., Task-specific monitoring of nuclear medicine technologists' radiation exposure, *Radiation Protection Dosimetry*, 109, 201–9, 2004.
11. Loevinger, R., Budinger, T. F., and Watson, E. E., *MIRD Primer for Absorbed Dose Calculations*. New York: Society of Nuclear Medicine, 1991.
12. Snyder, W. S., Ford, M. R., Warner, G. G., and Watson, S. B., *'S' Absorbed Dose per Unit Cumulated Activity for Selected Radionuclides and Organs*, MIRD Pamphlet 11. New York: Society of Nuclear Medicine, 1975. Also available online at http://interactive.snm.org/index.cfm?PageID=2199&RPID=1372.
13. International Commission on Radiological Protection, 1990 recommendations of the International Commission on Radiological Protection, ICRP Publication 60, *Annals of the ICRP*, 21 (1–3), 1991.
14. Clarke, R., 21st century challenges in radiation protection and shielding: Draft 2005 recommendations of ICRP, *Radiation Protection Dosimetry*, 115, 10–15, 2005.
15. Administration of Radioactive Substances Advisory Committee (ARSAC), *Notes for Guidance on the Clinical Administration of Radiopharmaceuticals and Use of Sealed Radioactive Sources*, March 2006. Available from the Department of Health website at www.advisorybodies.doh.gov.uk/arsac.
16. Smart, R. C., and Towson, J. E., Diagnostic reference activities for nuclear medicine procedures in Australia and New Zealand, *Radiation Protection in Australasia*, 17, 2–14, 2000.
17. Towson, J. E., Smart, R. C., and Rossleigh, M. A., Radiopharmaceutical activities administered for paediatric nuclear medicine procedures in Australia, *Australian and New Zealand Nuclear Medicine*, 32, 15–23, 2001.
18. Wagner, L. K., Lester, R. G., and Saldana, L. R., *Exposure of the Pregnant Patient to Diagnostic Radiations: A Guide to Medical Management*. Madison, WI: Medical Physics Publishing, 1997.
19. International Commission on Radiological Protection, Pregnancy and medical radiation, ICRP Publication 84, *Annals of the ICRP*, 30, 1, 2000.
20. International Commission on Radiological Protection, Radiation dose to patients from radiopharmaceuticals, ICRP Publication 80, *Annals of the ICRP*, 28, 3, 2000.
21. Russell, J. R., Stabin, M. G., and Sparks, R. B., Placental transfer of radiopharmaceuticals and dosimetry in pregnancy, *Health Physics*, 73, 747–55, 1997.
22. Russell, J. R., Stabin, M. G., Sparks, R. B., and Watson, E., Radiation absorbed dose to the embryo/fetus from radiopharmaceuticals, *Health Physics*, 73, 756–69, 1997.
23. Shi, C. Y., Xu, X. G., and Stabin, M. G., Specific absorbed fractions for internal photon emitters calculated for a tomographic model of a pregnant woman, *Health Physics*, 87, 507–11, 2004.
24. Mountford, P. J., Risk assessment of the nuclear medicine patient, *British Journal of Radiology*, 70, 671–84, 1997.
25. Mountford, P. J., and Steele, H. R., Fetal dose estimates and the ICRP abdominal dose limit for occupational exposure of pregnant staff to technetium-99m and iodine-131 patients, *European Journal of Nuclear Medicine*, 22, 1173–79, 1995.

26. Madsen, M. T., Anderson, J. A., Halama, J. R., Kleck, J., Simpkin, D. J., Votaw, J. R., Wendt, R. E., Williams, L. E., and Yester, M. V., AAPM Task Group 108: PET and PET/CT shielding requirements, *Medical Physics*, 33, 4–15, 2006.
27. International Commission on Radiological Protection, Release of patients after therapy with unsealed radionuclides, ICRP Publication 94, *Annals of the ICRP*, 34, 2, 2005.
28. Greaves, C. D., and Tindale, W. B., Radioiodine therapy: Care of the helpless patient and handling of the radioactive corpse, *Journal of Radiological Protection*, 21, 381–92, 2001.
29. Gillanders, J. A., Woods, S. D., and Jarritt, P. H., Radiation protection implications of a patient's death shortly after high activity radiopharmaceutical treatment, *Nuclear Medicine Communications*, 24, 341, 2003.
30. International Atomic Energy Agency, *Regulations for the Safe Transport of Radioactive Materials*, Safety Series 6. Vienna, Austria: IAEA, 1990.

8

Radiation Protection in External Beam Radiotherapy

Jim Cramb

CONTENTS

8.1 Introduction

External beam radiotherapy involves the delivery of high doses of radiation of up to about 80 Gy with the aim of curing a tumor by killing all the proliferating cells or of providing palliation by restricting the rate at which the tumor is spreading. To prevent adverse reactions from healthy tissue in the path of the radiation, radiotherapy is usually delivered with two or more beams each directed at the tumor from different angles, so that the total dose to the tumor is greater than that to any of the surrounding tissue. Also, tumor cells and normal tissue cells have a different capacity to recover from radiation damage. Radiation treatment aims to exploit this difference by delivering the radiation dose progressively over a period of up to 8 weeks, typically in 2 Gy fractions given five times per week.

Although external beam radiotherapy covers a range of radiation-emitting equipment, with differing characteristics and radiation protection requirements, the basic principles of room design and safety remain applicable in all cases. The following section provides a brief overview of the categories of equipment that might be found in a radiotherapy department, and subsequent sections give radiation protection and safety guidelines applicable to each of these types. Particular emphasis is given to the radiation safety of linear accelerators, because of the predominant role they play in modern radiotherapy.

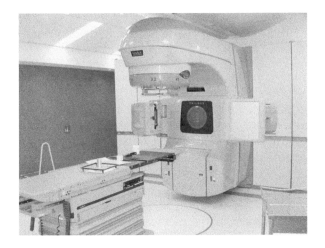

FIGURE 8.1
A linear accelerator and treatment couch.

8.2 Production of Radiation Beams Used for Radiotherapy

8.2.1 Linear Accelerators

Nowadays, the vast majority of external beam therapy is delivered with medical linear accelerators, so named because electrons are accelerated in a linear fashion along a waveguide. Figure 8.1 shows a modern linear accelerator.

Pulsed microwave power produced by a magnetron or klystron is transmitted to an accelerating waveguide. The timing of the pulses is tightly controlled and synchronized with bursts of electrons produced from the cathode (electron gun). The waveguide and bending tube are maintained under a high vacuum to prevent unwanted ionization as the electrons traverse the guide. The beam of electrons is then directed onto a metallic target (usually made of tungsten) to produce bremsstrahlung photons or onto a scattering foil to produce a wide electron beam. Single energy machines can produce x-rays up to 6 MV with a relatively short waveguide, of about 50 cm length, which can be mounted vertically in the treatment head. In linear accelerators designed to deliver higher-energy radiation the waveguide is mounted horizontally and the electron beam is turned through an angle by a bending magnet before being directed toward a target or scattering foil. Most linear accelerators of this type have a 270° bend to ensure tight energy discrimination. High-energy linear accelerators can produce two or even three photon energies (up to 25 MV) by altering the number of accelerating cavities in the waveguide and five or more electron energies. Rotating gantry and collimator systems allow beams to be directed at a tumor from different angles while the patient remains stationary on the treatment couch, and tertiary collimation (multileaf collimator or customized blocking) enables radiation to conform closely to the required treatment volume. The geometric location of the intersection of the axis of rotation of the gantry, collimator, and treatment couch is referred to as the isocenter. Figures 8.2 and 8.3 are schematic diagrams of a linear accelerator and treatment head, showing the key components.

Since electron beams are much less penetrating than photon beams, radiation protection considerations are governed by the absorption and scattering characteristics of high-energy photons. For beam energies above about 10 MV, photoneutrons are also produced; these can contribute significantly to radiation exposure and therefore need to be specifically considered in the room and shielding design.

8.2.2 Cobalt-60 Units

Prior to the development of the medical linear accelerator, cobalt-60 units were widely used for radiotherapy. They continue to be used in departments with limited resources, since they are cheaper and easier to maintain. The 1.1 MeV gamma rays from a cobalt-60 source are not as penetrating as the higher-energy photons produced by a linear accelerator, so room shielding requirements are generally less. However, unlike a machine that produces radiation only when energized, a high-activity radioactive source is a potential radiation hazard at all times, and hence additional safety precautions are required. Due to the 5.261-year half-life of 60-Co, the dose rate of cobalt

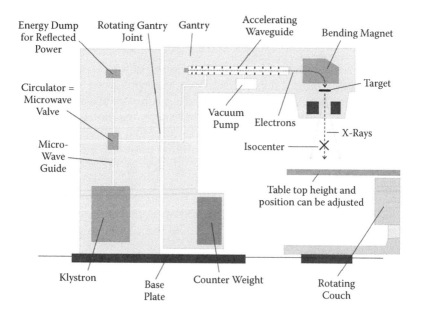

FIGURE 8.2

Schematic diagram showing the key components of a linear accelerator.

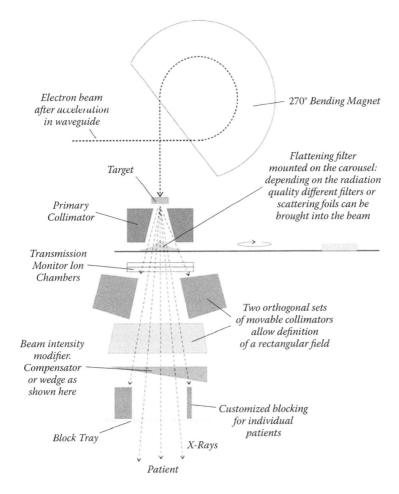

Electron beam
after acceleration
in waveguide

270° Bending Magnet

Target

Primary
Collimator

Flattening filter
mounted on the carousel:
depending on the radiation
quality different filters or
scattering foils can be
brought into the beam

Transmission
Monitor Ion
Chambers

Two orthogonal sets
of movable collimators
allow definition
of a rectangular field

Beam intensity
modifier.
Compensator
or wedge as
shown here

Block Tray

Customized blocking
for individual
patients

X-Rays

Patient

FIGURE 8.3
Cross section through the head of a linear accelerator in x-ray mode.

units reduces with time, and it is recommended to change the source at least every 5 years to ensure adequate dose rate, and therefore acceptable overall treatment times. The source change of a 60-cobalt unit is a complex procedure requiring specialized equipment and source containers. Consequently, it is associated with potentially high risk of exposure of staff. Typically the source changes are performed by company representatives.

8.2.3 Kilovoltage Therapy Units: "Superficial" and "Deep" Therapy

Machines utilizing conventional x-ray tube technology predate linear accelerators and are still in use in many radiotherapy departments for treating superficial (skin) tumors and for treatment of tumor masses and metastatic disease up to a few centimeters deep, the latter generally being for pallia-tion rather than cure. Superficial machines operate up to 100 kV, and deep machines up to 300 kV. Even in a large department, the workload of superfi-cial and deep machines tends to be quite low, since many of the former treat-ments can now be delivered using megavoltage electron beams. Therefore, a popular variation is a machine that combines the features of both, with a maximum operating potential of 225 kV.

8.2.4 Radiotherapy Simulators, CT Scanners, and CT Simulators

A radiotherapy simulator is a device that can replicate the movements of a linear accelerator and treatment couch, but is equipped with a conventional x-ray tube operating at up to 140 kV. It can be used both in a fluoroscopic mode to assist in determining the treatment area and in radiographic mode to provide a film or digital image on which the radiation oncologist can indi-cate the required treatment field size. This can then be used as a reference image to be compared with an image taken during treatment (portal image), as part of the process of verifying the patient's treatment.

Nowadays, the majority of radical radiotherapy treatments are planned using three-dimensional image sets, rather than planar images, and hence if a department has a simulator at all, it tends to be used mostly for pallia-tive and other relatively straightforward treatments. Simulators have been largely superseded by so-called computed tomography (CT) simulators (CT scanners designed specifically for radiotherapy use, with a wide bore so that patients can be scanned in the same position as for their treatment) or conventional diagnostic CT scanners (either within the radiotherapy depart-ment or shared with radiology).

The radiation protection issues for x-rays from simulators and CT scanners are covered in Chapter 6 of this book.

8.2.5 Tomotherapy

Tomotherapy is an emerging technology where a 6 MV beam from a linear accelerator is collimated to acquire a daily fan beam spiral CT of the patient, followed by radiotherapy, also using a fan beam that modulates as the gantry rotates and the couch moves longitudinally. The CT scan is used to fine-tune the radiotherapy from day to day. Compared to linear accelerators, the Hi-Art helical tomotherapy units have a relatively small footprint (see Fig-ure 8.4). The collimation to a fan beam geometry also reduces scatter, and the unit has imaging and shielding equipment at the opposite side of the linear accelerator on the ring gantry. Therefore, the unit comes with an "in built" beam stopper, reducing room shielding requirements.

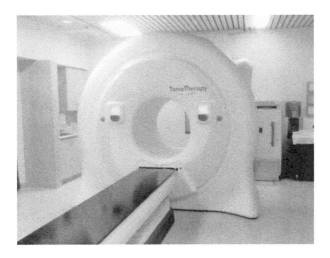

FIGURE 8.4
A tomotherapy unit in a treatment bunker.

8.2.6 Proton Therapy

Proton therapy has an advantage over gamma rays and x-rays when tumors are very close to critical structures. Protons deposit most of their energy in one region, known as the Bragg peak, which occurs at the point of greatest penetration of the protons in tissue. This depth is dependent on the energy of the proton beam, so by tightly controlling the energy, and using a range of energies, the Bragg peaks can sum to form a broader peak that falls within the tumor, as shown in Figure 8.5.

Neutrons are produced whenever protons are absorbed, so treatment rooms need to be designed specifically to minimize neutron exposure. Proton therapy machines are very expensive and do not offer any significant therapeutic advantage for the majority of tumors; hence there are only a few proton facilities throughout the world. Radiation protection and shielding requirements for proton facilities are not covered in this chapter.

8.3 Design of Treatment and Imaging Facilities

8.3.1 Exposure Limits and Constraints, Legislative Requirements

The International Commission on Radiological Protection (ICRP) specifies acceptable upper limits to annual exposure levels for radiation workers (as defined) and for nonradiation workers.[1] These are generally accepted worldwide and guide the radiation safety legislation of countries, states, and

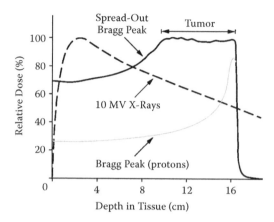

FIGURE 8.5
Showing the advantage of proton beams over a 10 MV x-ray beam.

provinces. Despite this, there are significant variations in the way these limits are interpreted from country to country and even between states or provinces in the same country. For instance, in many countries, emphasis is given to the maximum instantaneous exposure rate and the maximum exposure per hour, as well as the annual exposure. A controversial issue at present is whether a facility should be designed so that members of the public can potentially receive up to the maximum allowable exposure from that facility alone (i.e., the exposure constraint equals the exposure limit), or whether there should be a safety factor to allow for possible exposures from other radiation sources. In this chapter we deal with broad principles, rather than debating such issues or attempting to cover specific jurisdictional requirements. It is important, therefore, that the serious reader also acquaint himself or herself with the local regulatory requirements pertaining to a treatment facility. It may be necessary to submit a detailed plan including predicted exposure levels before permission to construct a new facility is granted.

The underlying principle behind the design of any room housing radiation-emitting equipment is that the exposure to all staff, patients, and the general public must follow the ALARA (as low as reasonably achievable) principle for all possible operating conditions of the equipment. In practice, this means that calculated annual exposure levels and measured exposures and exposure rates are very comfortably below regulatory requirements, even by up to an order of magnitude where possible.

8.3.2   Linear Accelerator Bunkers

The predominant interaction process for megavoltage photons is Compton scattering (see Chapter 2), whereby an outer shell electron is ejected and a

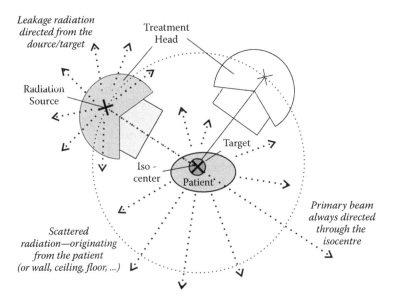

FIGURE 8.6
Sources of radiation around a megavoltage treatment unit.

photon of reduced energy is produced, deflected at an angle to the incident photon. For higher energies (above about 10 MV), pair production (see Chapter 2) becomes progressively more significant. Electrons and positrons then recombine to produce photons of energy 511 kV, which in turn are most likely to undergo Compton scattering. Thus, in general multiple scattering events take place before the initial photon energy is dissipated, and this is reflected in the design of a linear accelerator or megavoltage treatment room, which is commonly referred to as a bunker.

The bunker needs to be designed to restrict three types of radiation, which are illustrated in Figure 8.6:

1. Primary—the direct beam
2. Scatter—radiation that has scattered from the treatment head, accessories, patient, bed, walls, floor, or anything else in the treatment room
3. Leakage—radiation that is emitted from the equipment other than through the defined collimation system

This section sets out to give an introduction to bunker design and some practical advice with some simplified sample calculations. For additional information, refer to references 2–5, which deal specifically with bunker design.

Because radiation intensity reduces according to the inverse square law, it is clearly an advantage to have a large bunker. A roomy bunker also creates a more pleasant environment for staff and patients, the only downside being the extra distance traveled walking from the control area to the treatment couch. No particular design has emerged as optimal, however. Bunkers come in a wide variety of geometries and sizes, reflecting constraints of space and location, individual design philosophies, and specific local functional and aesthetic preferences.

It is important to consider if there are any special treatments or other services to be performed. One such example is intensity-modulated radiotherapy (IMRT), which is discussed in Section 8.7.4. Another is total body irradiation (TBI), which requires more space. Research work and uses such as blood irradiation for sterilization may alter the use pattern of the linac. Other issues that may be considered in bunker design are operational aspects, such as ease of visualization of the entire bunker (to ensure it has been vacated) and storage space.

For megavoltage photons undergoing Compton scattering, the rate of absorption is quite similar for various materials, allowing for their mass density. This means that the choice of material for a bunker can be based primarily on practical considerations such as cost, availability, and ease of construction. The vast majority of bunkers are constructed of concrete, of density 2,350 kg per cubic meter. There are other options, however, including high-density blocks such as Ledite®, which are self-supporting and can be stacked together to form a compact (and semiportable) bunker.

8.3.2.1 Primary Barriers

Primary beam barriers need to be thick enough to reduce the beam intensity by several orders of magnitude. Clearly, the primary barrier needs to be much more attenuating than the barriers for scatter and leakage—typically it will be at least twice as thick. Therefore, if the bunker is made entirely of concrete, the primary barrier will either protrude into the room, taking up valuable space, or jut out from the wall on the outside of the bunker. To avoid this, sometimes a steel plate is incorporated into the primary barrier. Steel is approximately four times as dense as concrete, so a combination of steel and concrete in the right proportion allows the primary barrier to match the other walls in thickness. Steel does add significantly to the cost, however. Another option when space is tight is to use an aggregate with a higher metallic ore content to produce high-density concrete, with a density of 3,500 kg per cubic meter. Finding a contractor who will guarantee this high density can be difficult, however. Lead sheet is widely used in kilovolt installations, in both radiology and radiotherapy. Since lead is approximately six times as dense as concrete, it might seem to be a logical choice for megavolt therapy as well, to keep the walls as thin as possible. However, lead is impractical to work with when thicknesses of more than a few millimeters are required. It is expensive, not self-supporting, and may creep over time due to gravity. Ledite blocks could also be used for part of the primary barrier.

The gantry of most accelerators is designed to rotate through a full 360°; therefore, the primary beam is contained within a solid angle striking the central section of two of the walls, the floor, and roof. The primary barriers must be wide enough to be sure of intercepting the largest possible field size. Allowance should be made for the fact that the collimator system of a linear accelerator can also rotate. The maximum collimator setting of most accelerators is 40 × 40 cm, so the primary barrier should have a width, W_P, of at least

$$W_P = \frac{d_{SX}\sqrt{0.4^2 + 0.4^2}}{d_{SI}} \tag{8.1}$$

where d_{SX} is the distance from the source to the exit side of the barrier (6 m in Figure 8.6) and d_{SI} is the distance from the source to the isocenter. It is common practice to allow an additional margin on either side for good measure. This is not strictly necessary, since the x-ray beam is initially shaped by a fixed conical primary collimator to a diameter several centimeters or so less (at the isocenter) than the diagonal of the maximum square field, so the maximum field size is clipped at the corners. However, an extra margin means the primary barrier will be able to intercept small-angle Compton scatter, which predominates at high energies, and also allows for a little flexibility in positioning when the accelerator or future accelerators are installed.

A typical bunker has laser units fitted on the side walls, ceiling, and end wall to project sagittal, coronal, and transverse laser lines at the isocenter to facilitate accurate patient positioning. The lasers must be firmly mounted on a solid surface, not on the finished walls, and it is common for the sake of appearance for the side laser units to be recessed into the primary barrier walls. Since these lasers are at the height of the accelerator's isocenter, they compromise the barrier thickness in the very region that will most often be irradiated. Therefore, if the primary barrier is made of concrete, a sufficient thickness of steel plate should be added behind the laser to compensate, using a 4-to-1 rule of thumb. For example, if the laser is to be recessed to a depth of 15 cm, the recess needs to be 20 cm deep, with a 5 cm steel plate behind the laser.

From simple geometry the diverging beam is considerably wider when projected to the bunker walls than onto the ceiling directly above, and similarly, the oblique path length through the roof increases as the gantry is moved away from the vertical. Oblique beam directions also tend to be used less frequently. Therefore, in principle, the roof could be tapered or stepped in both width and thickness, reducing construction costs. In practice, though, construction is simpler when rectangular shapes are formed up and poured.

It is an advantage to construct bunkers in the lowest floor of a building, so that the floor does not need to be considered when calculating barrier thicknesses. Also, it is an advantage when two or more bunkers are designed and built together to share a common primary barrier. If adjacent areas and the space above are unoccupied or are only very occasionally occupied, barriers can be correspondingly thinner, although care must be taken that the occupancy does not change at a later date.

A special case arises when a bunker is part of a single-story facility, with no possibility of any future construction above. Clearly, both the primary barrier and the rest of the roof can be much thinner than would be necessary for an occupied area. However, if the roof is too thin, "sky shine" (radiation scattered back toward the ground from air above the roof) may become significant. Also, consideration must be given to future developments: Is there a possibility of an adjacent multistory building? If the bunker roof thickness does not allow for any occupancy above, appropriate signage or barriers must be erected, to prevent tradesmen from ever going onto the roof while treatments are in progress.

8.3.2.2 Secondary Barriers, Mazes, and Bunker Geometry

Scatter and leakage radiation are often grouped together as secondary radiation. The bunker needs to be designed in such a way that the amount of secondary radiation exiting the bunker is low. Therefore, transmission through the remaining walls and roof (and floor if relevant), as well as the amount of scattered radiation reaching the bunker entrance, need to be considered. Neutrons produced from photons of energy above 10 MV are also secondary radiation, but are scattered and absorbed quite differently from photons and must be considered separately when designing the bunker.

The total contribution of secondary radiation in any one direction is two or three orders of magnitude less than the primary fluence at the same distance from the isocenter. For megavoltage beams, the angular distribution of Compton scattered photons is peaked in the forward direction, increasingly so with increasing energy. From Table B4 of NCRP 151[3] it can be seen that for a human-sized phantom and a field size of 400 cm², the scatter fraction for 18 MV photons at 10° is 1.42×10^{-2}, while at 90° it is only 1.89×10^{-4}, almost two orders of magnitude less. This angular dependence can be exploited in the design of a bunker. However, a conservative and straightforward approach is to consider that the scatter in all directions is the value at 10°—approximately 1% for 6 MV and 1.5% for 18 MV treatments.

Bunkers are commonly accessed via a maze, as shown in Figure 8.6. The maze must be designed in such a way that radiation cannot reach the entrance without being scattered at least twice. The use of a nib, or wall extending from the maze a short distance into the bunker, can make a significant difference to the dose at the maze entrance by greatly reducing the solid angle for scattering along the maze. Similarly, an extra bend at or near the entrance of a maze is an advantage. The photon dose at the maze entrance can be approximated by considering (1) scatter from the patient and floor, (2) scatter from the region of wall near the maze, and (3) scatter partway along the maze, each time also reduced by the inverse square law over the path lengths $d1$, $d2$, and $d3$. Depending on the bunker design, multiple scatter from the primary barrier on one wall, weighted according to the proportion of time the beam points in that direction, should be added, as well as leakage radiation. Monte Carlo simulations can be used for such purposes.[6] Even these calculations, however, can still only be considered as approximate and must be confirmed by radiation survey measurements once the accelerator is installed.

If there is insufficient floor space available, there may be no option but to have a short maze together with a heavy door. Sometimes, a door even leads directly into the bunker. It is a challenge to design a shielding door that is sufficiently thick but efficient and reliable to operate. Particular care must be taken to ensure that there is not excessive radiation at the edges of such a door.

Radiation protection doors are usually motor driven, but there must be an efficient backup manual method of opening the door in the event of mechanical failure. It is also important to ensure no one can be squashed by the door, and a pressure mat is typically installed to stop door motion once someone enters the path of the door.

8.3.2.3 Neutrons

Doors designed to absorb just photons are relatively rare; high-energy linear accelerator bunkers are more likely to require a door for neutrons. For beam energies above about 10 MV, unwanted neutrons are produced as x-rays interact with materials in their path, particularly the collimators and beam accessories, but also the patient, floor, and walls. The cross section for (photon, n) interactions increases significantly with increasing energy. Therefore, for beams of 18 MV and above, the equivalent dose from neutrons at the entrance of a simple maze will probably be greater than the photon dose, unless material with a high cross section for neutron absorption is added.

Neutrons scatter many times before being brought to rest. Any material with a large proportion of hydrogen atoms is a reasonably good neutron absorber; therefore, a concrete bunker and maze walls will be sufficient provided the maze is long enough. However, boron- or lithium impregnated polyethylene has a particularly high cross section for the absorption of neutrons. Since the neutron dose at the maze entrance is related to the cross-sectional area of the maze, one strategy is to suspend a baffle of polyethylene or any other hydrogenous material above head height in the maze. The baffle is more effective if it is placed at or near the bunker end of the maze. Another option that can be effective without compromising the bunker's aesthetic appearance is to line the maze and bunker walls with boron-impregnated polyethylene, particularly the corner where the maze meets the bunker.

There are varying views on the calculation of neutron doses. Typically, empirical relationships are used, rather than attempting to model the physical processes. For example, Kersey[7] has published equations that predict the neutron intensity at the maze entrance, as a function of the path lengths in the bunker and along the maze, and the cross-sectional area of the maze. He assumes that the neutron dose will reduce by a factor of ten for every 5 m of maze length. McGinley and Butker[8] have shown that this agrees only moderately well with measurements. They also found that a right-angle turn in the maze reduces the dose by a factor of three relative to a straight maze of the same length.

A typical neutron door comprises sheets of boron-impregnated polyethylene and lead, sandwiched between steel plates—the lead and steel serve to absorb the gamma rays from (n, γ) absorption, and the steel gives the door strength and rigidity. Sliding or swinging doors can be used, and they can be hydraulically or mechanically operated. It is not as important for a neutron door to have no gaps, because the neutron fluence is much more isotropic in direction, and few neutrons will pass directly through any gap.

8.3.2.4 Other Considerations

An access channel or channels need to be provided through the bunker wall for the supply of power and water and for communication and signal lines between the linear accelerator and control computers outside. To minimize the escape of secondary radiation, these channels should be below floor level (so that the walls are intact above floor level) and no larger than necessary. It is acceptable to include a narrow (approximately 10 cm diameter) duct through the wall above floor height to accommodate the cables used for physics measuring equipment, provided that the duct is angled rather than horizontal. For example, the duct could be just above bench height on the outside of the bunker and at floor height on the inside. Another solution to minimize radiation exposure is to have another duct below floor level through which physics cables are permanently threaded, or (safest of all but not as convenient) simply ensure that all physics cables are long enough to be run into the bunker via the maze. Above ceiling height, there will also need to be an HVAC (heating, ventilation, and air conditioning) penetration. This is unlikely to be a problem unless the space above is occupied, in which case the opening must be well away from a primary beam direction.

When a new bunker is constructed, it is important to supervise the pouring of the concrete walls, in particular the primary barriers. Often, abutting sections are formed and poured individually. The particular technique used must not have any joins in critical places that might allow excessive transmission. If joins in the primary barrier are unavoidable, it is best if the total thickness is poured in stages, with the joins staggered. It helps if the joins can also be angled relative to the direction of the primary beam.

The linear accelerator must be interlocked in such a way that it cannot be turned on if a shielding door (if required) is not closed. For low-energy beams that do not require the door to be closed, and bunkers with no door, there must be some other means of ensuring nobody enters the room while the beam is on. An interlocked gate is the most common, but other devices such as a laser or infrared beam across the maze entrance can be just as effective providing all staff are well trained in their purpose. The gate must be closed before beaming can commence, and opening the gate must cause beaming to cease. It is also good practice, and mandatory in some jurisdictions, to have a "last person out" button within the bunker, which is pressed as the room is vacated by staff. The button should be positioned where it is easy to see that there is in fact nobody else (other than the patient) still in the room. The interlocking circuitry should include a timer so that the gate must subsequently be closed within, say, 10 s. It should also emit a clearly audible and recognizable sound, so that even if the person pressing the button has not actually checked that the room is vacant, anyone remaining will be alerted.

It is also a universal requirement for a "beam on" light to be illuminated when radiation is produced. It is generally acceptable for this light to be coupled to the linear accelerator circuitry, but sometimes regulations necessitate that it be coupled to an independent in-room radiation detector, as is required for a cobalt-60 bunker or brachytherapy room. Although it is quite safe for nonradiation workers to be in the control area of a linear accelerator from time to time, the facility design should be such that access is limited; that is, the control area should not be near a thoroughfare, and there should be signage, doors, or both to discourage close approach. Some, but not all, jurisdictions require that the trefoil "Danger—Radiation" sign be permanently displayed outside linear accelerator bunkers.

8.3.2.5 Design Constraints: Calculating Barrier Thicknesses

To determine appropriate thicknesses of primary and secondary barriers from first principles, the standard approach is as follows:

Define which of the areas adjacent to the bunker, and above and below if appropriate, are supervised and which are uncontrolled. For example, the spaces used by the machine operators, nearby patient waiting areas, and patient changing and preparation rooms are supervised areas. The person in charge of this area will be a trained radiation worker who can ensure that safe practices are followed. Staff who work routinely in a supervised area are designated as radiation workers, who, in many jurisdictions, have an annual allowable radiation exposure of 20 mSv in accordance with the 1990 ICRP recommendations.[1]

Uncontrolled areas, as the name implies, may be occupied by other staff and members of the public, without the knowledge of the operating staff, and conversely perhaps without them being aware that they are in the proximity of radiation-emitting equipment. Examples are clerical staff on the floor above, or pedestrians on the pavement just outside the building. In many jurisdictions the allowable annual exposure in an uncontrolled area is therefore that of the general public, 1 mSv.

Every facility should be designed so that radiation levels follow ALARA principles, not just on the borderline of acceptability. Of course, ALARA principles mean that cost and space do have to be considered as well. Jurisdictional radiation safety regulations may prescribe maximum annual exposures that are lower than those of the 1990 ICRP recommendations. As well, it is not uncommon to specify a maximum instantaneous exposure rate or rate per hour or rate per week. A design constraint that nobody in a controlled area is to receive more than one-tenth of the allowable exposure (Hlimit = 2 mSv per annum) is usually quite achievable. Uncontrolled areas, however, are not as straightforward. As alluded to in Section 8.3.1, some countries have recently introduced regulations that require an Hlimit of only 0.3 mSv from any one radiation source in an uncontrolled area. This has been a point of much discussion because to many it seems unreasonably low for a potentially fully occupied area, and in many circumstances it is also at the limit of detection for dosimetry. It is perhaps more reasonable if used together with a realistic estimate of the amount of time that one person could spend in each uncontrolled area.

8.3.2.5.1 Occupancy Factor (T)

An occupancy factor needs to be assigned to each area in the proximity of the bunker. Occupancy factors given in NCRP Report 151[3] are:

T = 1 for all controlled and supervised areas (which includes the entire area occupied by radiation workers while on duty) and for adjacent fully occupied areas such as offices, wards, and attended waiting rooms.

T = 1/5 for uncontrolled areas such as corridors, rest rooms, and utility rooms.

T = 1/20 for public toilets, storage areas, and unattended waiting rooms.

T = 1/40 for outdoor areas and other areas that are only occasionally occupied, such as stairways, pavements, and unattended parking lots.

Previously published occupancy factors were extremely conservative. Even the above factors are conservative—for example, a "fully occupied" area is unlikely to be occupied all the time by the same person, and nobody is likely to occupy a stairway for an hour a week, every week of the year. Lower factors can be used when designing a facility, provided it is certain that the factors will be applicable both now and in the future.

One location where a considerably lower factor could sensibly be applied is immediately adjacent to the bunker maze entrance. Although strictly this has a T value of 1 by definition, radiation workers can readily be instructed not to stand near the entrance. In any case, it is physically impossible to operate the machine console and stand near the entrance at the same time.

8.3.2.5.2 Use Factor (U)

A measured or calculated exposure at any point applies only to one particular machine geometry, for example, with the gantry pointing at one wall. Consideration needs to be given to the fraction of time that the machine is in this position. Again, it is best to be conservative, particularly for the primary barriers. For example, factors of downward 0.5, upwards 0.5, and horizontally 0.25 each side give a sum of 1.5 rather than 1.0, which seems illogical. However, they represent an upper limit on the use in any direction and provide a safeguard for the future, if it is decided to use a machine just for particular types of treatment. A classic example is total body irradiation, which typically entails beaming against one wall, with the collimators at their maximum setting. The proportion of higher and lower photon beam energies should also be incorporated into the use factors. Radiation exposures will be highest for the highest-beam energy, so the high-energy use factor should be generously overestimated.

8.3.2.5.3 Annual Workload (W)

The annual workload (total output), W, of the machine needs to be estimated. This can be done based on both past experience of patient throughput and projected working hours. For example, suppose the machine is going to be operated 250 days per year, 10 h per day, 4 patients per hour with each patient receiving a 2 Gy treatment. Machines are usually calibrated so that 1 monitor unit (MU) corresponds to 1 cGy under reference conditions. Typically, a total of about 300 MU is required to deliver 2 Gy, so the projected MU per year is $250 \times 10 \times 4 \times 300 = 3 \times 10^6$, which corresponds to $W = 3 \times 10^7$ mSv at the isocenter.

The above calculation does not take into account morning warm-ups, calibrations and quality assurance checks, and experimental or developmental work. These should also be included in an estimate of W. It is reasonable to assume that the beam is always directed downwards for these types of irradiations.

The primary component of radiation essentially obeys the inverse square law (ISL) from the point of production. Most linear accelerators have a source to isocenter distance of 1 m; therefore, in the absence of any absorption, the dose rate at a point, say, 5 m further on than the isocenter will have fallen to $(1/6)^2 = 1/36$.

The above factors can be incorporated into the following equations:

$$Hprim = W \times ISLp \times U \times T \qquad (8.2)$$

$$Aprim = Hlimit/Hprim \qquad (8.3)$$

where Aprim is the required attenuation of the primary barrier.

Generally, no allowance is made for absorption within the patient or test phantoms, to give a conservatively low value for Aprim.

This is often quoted as a number of tenth value layers (TVLs), where one TVL is the thickness required to reduce the intensity of the beam to one-tenth.

A value of 0.00001 for Aprim therefore corresponds to five TVLs.

Now the required barrier thickness in centimeters can be determined. Transmission data for concrete, steel, and lead as a function of beam energy and wall thickness are given in references 3 and 4. As well, most linear accelerator manufacturers will supply an installation package, giving suggested layouts, recommended wall thicknesses, and transmission data sheets for their particular beam energies. However a useful and conservative rule of thumb is 1 TVL = 50 cm of concrete. This is about right for the high-energy beam of most linear accelerators (15–25 MV) and an overestimate for lower-energy beams.

If Aprim requires five TVLs, then the primary barrier will need to be 5 × 50 cm = 250 cm thick.

The determination of the required thicknesses of the secondary barriers follows the same methodology as for the primary. However, as discussed above, a rigorous calculation for secondary radiation is much more difficult. One simplification is to use a U value of 1, even though the scatter and leakage reaching any point do vary somewhat with gantry angle. Another approximation is to assume that all scatter comes from the patient or from directly above or below. Then

$$\text{Hsec} = W \times \text{ISLs} \times T \times \{f\ s(\varphi, \theta) + fl(\varphi, \theta)\} \tag{8.4}$$

$$\text{Asec} = \text{Hlimit}/\text{Hsecond} \tag{8.5}$$

ISLs apply to the origin of the scatter, which can be approximated as the isocenter; fs is the amount of scatter in a particular direction, as a fraction of the primary beam intensity; and fl is the amount of leakage radiation in a particular direction, as a fraction of the primary beam intensity.

The above equations will yield different values for Asecond at different positions around the walls. However, as discussed above, a further simplification is to ignore the angular dependence of fs and fl, and use fs = fsmax (the maximum value at any angle, from tabulated data), fl = flmax (must be less than 0.001 by regulatory requirements).

Usually, to achieve the same Hlimit, the secondary barrier can be at least two TVLs thinner than the primary barrier. The mean energy of secondary radiation is less than for the primary beam, so the TVL is also less; however, the most conservative approach is again to use a TVL of 50 cm of concrete. Typically, secondary barriers are 100 to 150 cm thick, depending on the occupancy factor outside.

For the bunker shown in Figure 8.7, suppose that the workload is W = 3 × 10^7 mSv at the isocenter and that 18 MV photons will be used for all treatments. The room on the right of the bunker is a utility room, which is used by nonradiation workers; therefore, Hlimit = 1 mSv and T = 1/5.

Consider point A: For the primary beam barrier, U = 0.25:

The distance from the source to outside of the wall is 6 m; therefore, ISLp = 1/36. Thus,

Hprim = W × ISLp × U × T = 3 × 10^7 × 1/36 × 0.25 × 1/5 mSv.

Hprim = 4.17 × 10^4 mSv.

Aprim = 2.40 × 10^{-5}.

No. of TVLs = 4.6.

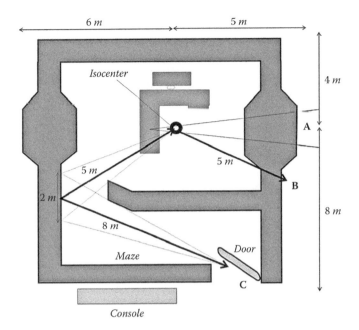

FIGURE 8.7
A possible bunker design, in a situation where space is fairly limited.

Therefore, the primary barrier needs to be 230 cm thick.

Consider point B: For a secondary barrier, if we ignore any dependence on gantry angle, U = 1:

Assume all scatter comes from the patient; the distance from the isocenter to the outside wall is 5 m; therefore, ISLs = 1/25.

Assume an average field size of 400 cm².

Assume fs(φ, θ) + fl(φ, θ)} = fsmax + flmax = 0.015 + 0.001 = 0.016.

Therefore, Hsec = W × ISLs × U × T × f = 3 × 10⁷ × 1/25 × 1 × 1/5 ×.016 mSv.

Hsec = 3.84 × 10³ mSv.

Asec = 2.60 × 10⁻⁴.

No. of TVLs = 3.6.

Therefore, the secondary barrier would need to be 180 cm thick.

In this case, a simplified calculation produces a value that is much thicker than is required in practice. As well as assuming that only the highest energy beam is used, the strong angular dependence on scatter has been ignored, the average field size is generous, and the TVL for scattered radiation has been rounded up.

Let us see what difference the application of use factors makes:

U = 0.25 beaming toward the utility room, fs = fsmax = 0.015, fl = flmax = 0.001.

U = 0.50 beaming up or down, fs at 90° = 0.00019, fl = flmax = 0.001.

U= 0.25 beaming away from the utility room, fs at 180° < 0.00012, fl = flmax = 0.001.

Therefore, Hsec = W × ISLs × T × {Σ U(θ) × f(θ)} = 3 × 10^7 × 1/25 × 1/5 × (0.25 × 0.016 + 0.5 × 0.00119 + 0.25 × 0.00112) mSv.

Hsec = 1.17 × 10^3 mSv.

Asec = 8.55 × 10^{-4}.

No. of TVLs = 3.1.

Therefore, the secondary barrier needs to be only 155 cm thick.

This is still conservative, since the TVL for radiation scattered through larger angles is much less than 50 cm.

For a high-energy beam, the photon dose at the maze entrance is a combination of secondary radiation (mostly scattered from the patient, but also scattered from the primary barriers, and leakage) and photons produced by the absorption of neutrons along the maze. A full calculation of photon dose is very complex and is not included in this chapter. In any case, reduction of neutron dose will be of greater concern when designing a door.

To illustrate the principles behind calculations involving multiple scatter, we will now consider point C at the maze entrance, assume a beam energy of 6 MV, and consider only photons scattered from the patient (which is the major contributor to dose at the maze entrance for a 6 MV beam). From Figure 8.7, the scattered radiation traverses a distance of 5 m from the isocenter to the left wall, then is reflected a further 8 m to the entrance. The reflection coefficient α for this geometry is approximately 2 × 10^{-2}, assuming a mean scattered energy of 0.5 MeV[3] (Tables B8a and B8b).[3]

W = 3 × 10^7 mSv as before.

U = 1, T = 1, Hlimit = 2 mSv.

Take fs(φ, θ) to be fsmax, 0.010.

A is the wall area that can be seen from both the isocenter and the maze entrance—in this case, 2 m wide by 3 m high, that is, 6 m^2.

Therefore, Hentr = W × ISLs(1) ISLs(2) × U × T × fsmax × α × A

= 3 × 10^7 × 1/25 × 1/64 × 1 × 1 × 0.010 × 0.020 × 6 mSv.

= 22.5 mSv.

Aentr = 2/22.5 = 8.9×10^{-2}.

Therefore, based on the above calculation, a door to the maze, allowing transmission of less than about 9%, is needed. The door would require several centimeters of lead or steel.

(A more detailed calculation of patient scatter, considering the scattering angle fs as a function of gantry angle, would result in a thinner door. In addition, as discussed above, it is not unreasonable to assign a value of T < 1 for a position such as C, which will only be very rarely occupied in practice.)

8.3.3   Kilovoltage, CT Simulator, and CT Rooms

The walls of the rooms housing kilovoltage units are commonly lined with lead sheet. However, a solid concrete or brick structural wall will usually provide adequate attenuation of the beam, without the need for extra shielding. A similar methodology as used for megavoltage beams can be applied to the calculation of barrier thicknesses, the main difference being that secondary radiation is of less concern. This is because kilovoltage beams are predominantly absorbed by photoelectric interactions, and the range of secondary electrons is very short. At higher kilovolt settings, Compton scattered photons start to become significant. Data on transmission of kilovolt beams through lead and concrete are contained in NCRP Report 147.[9] Typically, wall thicknesses equivalent to 2 mm of lead are adequate, when appropriate use and workload factors are assigned.

A conservatively designed treatment room will have all walls fully shielded, although this is not strictly necessary. Kilovoltage machines are usually capable of pointing in any direction, but the nature of kilovolt therapy is such that the beam is generally angled either directly or obliquely downwards, using low-kilovolt (soft) beams that do not penetrate through the patient. Beaming in directions where there is inadequate shielding can therefore be restricted, either physically by limiting machine movements (safest) or by signage and specific operating instructions. Figure 8.8 illustrates this point.

If lead lining is used, care must be taken to ensure that there are no gaps between lead sheets and that lead is adequately fitted around plumbing fittings and other wall penetrations. The room will often be fitted with two doors that should also be lined with lead. Sliding doors are preferable, since they can overlap the wall when closed. Both doors should be interlocked so that the machine cannot be turned on unless the doors are closed, and so that the exposure terminates as soon as a door is opened. The interlocks should be checked regularly for functionality.

It is essential that patients can be observed throughout their treatment. Rooms have traditionally included a lead glass or even a plain glass window of sufficient thickness; however, a camera and monitor are a good, and probably cheaper, alternative.

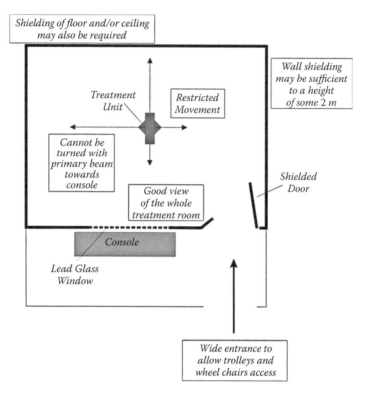

FIGURE 8.8
One possible layout of a kilovoltage therapy room.

The design considerations for a radiotherapy simulator are similar to those for a kilovolt treatment room. One difference is that the beam direction is restricted to rotation about the gantry axis, like a linear accelerator, so that only two walls can be directly irradiated. Although rigorous calculations of required barrier thicknesses could be done, a simulator can be used in a variety of ways, and it is harder to estimate appropriate use and workload factors. Shielding the room throughout with 2 mm of lead, or equivalent, will be more than adequate, and allows for flexible use of the room in the future.

CT scanner rooms pose less of a radiation hazard. The x-ray beam is finely collimated and is almost completely absorbed within the patient and CT gantry; therefore, exposure outside the room is largely due to scattered radiation. If the room is of moderate size and the gantry is remote from the operating console, the exposure to the operator will be well under regulatory limits, even without any specific shielding. However, following the ALARA principle, it is best practice to line the room with a lead sheet and fit interlocking lead-lined doors, as for a simulator or radiology room. For a more detailed discussion on CT scanner facilities, see Chapter 6.

8.3.4   Tomotherapy Bunkers

Design principles for a treatment bunker for helical tomotherapy (HT) units are similar to those for a conventional 6 MV linear accelerator, although consideration must be given to the longer beam-on times resulting from the use of small apertures, and hence greater proportion of secondary radiation, mainly due to increased leakage.

An HT unit does not employ a flattening filter, and as such the output is higher than a conventional linear accelerator. Under these circumstances it is useful to consider workload in terms of beam-on time, not number of monitor units delivered at the axis of rotation.

As discussed above, primary radiation is of limited concern as the beam is collimated to fan beam geometry and directed onto a shielded imaging device on the opposing side of the ring gantry. The very need to balance the ring gantry requires ample material at the location of the exit side of the beam that provides adequate shielding of the primary beam. A summary of design considerations for a helical tomotherapy unit has recently been provided by Zacarias et al.[10]

8.4   Radiation Surveys

As soon as possible after a new linear accelerator is capable of producing radiation, a detailed radiation survey should be conducted. This is especially important if the bunker is also new, because this will be the first opportunity to check whether the design constraints have been satisfied—both the design assumptions and the construction are being tested. It is best if the machine's output for the reference conditions for each photon energy is set to be at least approximately correct prior to the survey, to avoid misleading survey figures.

Equipment required includes the following:

- Two people, one to operate the linear accelerator and one to measure, with means of communication
- A photon survey meter (see Chapter 4), suitable for the energy range of the radiation measured, and with a calibration figure traceable to a standards laboratory
- If photon energies above 10 MV are to be measured, a neutron meter that reads directly in millisievert and millisievert per hour (see Chapter 4) and which has a reliable calibration figure
- A plan of the bunker and surrounds (including the floors above and below if appropriate)
- A tape measure

If parts of the survey require access to occupied areas, it is prudent to arrange to conduct the survey after-hours. If this cannot be done (e.g., for a hospital ward), then be sure to liaise with the appropriate people regarding the timing of and reason for the survey. Particular care should be exercised with a new facility. When timelines are tight, the survey might have to be undertaken while building works continue around and sometimes within the bunker, and before all the standard safety features have been installed. On the positive side, performing a survey at this stage often allows for speedy and easy fixing of identified issues such as unshielded cable ducts or air vents that compromise the integrity of the treatment room. Close liaison and good communication with all personnel on site is essential, for both radiation safety and good industrial relations.

Determine beforehand the appropriate points for survey readings to be taken, and for what beam geometries, concentrating on areas where the readings are likely to be highest, according to the design. The doorway or entrance to the maze, areas behind or near the primary barriers, and areas around wall penetrations are the most important. The surveyor should not be restricted to the predetermined points, though, as the measurements may reveal other points that warrant attention. When measuring the exposure through a wall or door, it is usual to hold the survey meter approximately 30 cm away from the barrier, at a height of about 1 m. This can be taken as an upper limit of exposure to a person standing or sitting in that vicinity. Of course, other heights and distances from the wall can also be measured, considering children and tall persons, as appropriate. Apply common sense; for instance, in the treatment console area, the workbench may prevent close approach to the wall.

Readings behind primary barriers should be taken with the collimators set to their maximum, with the collimator rotated through 45°, and with no scattering material in the path of the beam. Conversely, readings through secondary barriers and at the maze entrance should be taken with a "full scatter" phantom (at least 30 cm³) in the beam. A thorough survey will include measurements at gantry angles 0°, 90°, 180°, and 270°; however, the exposure through secondary barriers is not strongly dependent on gantry angle, so one angle will suffice where levels are very low. Depending on the bunker design, photon and neutron readings at the maze entrance are likely be highest when the gantry is pointing in the general direction of the maze, that is, gantry 90° or 270°. However, unlike photons, the neutron exposure is likely to be higher when the photon collimators are closed down. To establish this, it is best to measure for small, medium, and large field sizes at the maze entrance, then thereafter use the field size that gave the maximum neutron exposure.

Survey exposure rates should be entered into a spreadsheet and processed along with workload, use, and occupancy factors before establishing whether the facility is safe for long-term use. If a weakness in the design or construction is detected such that exposure is unacceptably high or borderline in some regions, act promptly to redress the situation. First double-check the measurements and calculations, then discuss possible solutions with the project manager. These might include thickening a wall with high-density blocks or steel plates, adding a door, converting an area from full occupancy to partial occupancy, or placing warning signs or a barrier so that an area can no longer be occupied at all.

Many jurisdictions require a formal detailed radiation survey report to be submitted before the equipment can be registered for ongoing operation.

8.5 Other Acceptance Tests Relating to Safety

8.5.1 Head Leakage

The radiation due to head leakage is expressed as a percentage of the reading at the isocenter for a 10 × 10 cm field. The International Electrotechnical Commission (IEC)[11] specifies that head leakage should be:

1. Less than 0.1% for a circle of 1 m from the target, in a plane perpendicular to the beam axis through the isocenter (excluding, of course, from the region of the treatment area)
2. Less than 0.1% for all points away from the patient area, perpendicularly 1 m from the path of the electron beam as it travels through the accelerator

Although most makes of accelerators meet these requirements quite comfortably and can provide typical leakage figures for their machines, it is still important to check each individual machine in case any of the shielding surrounding the waveguide or head has been omitted, or in case of gross electron beam steering errors. Discrete measurements at selected points will not necessarily detect a small area of high leakage, so it is good practice to completely wrap the treatment head in ready-packed radiotherapy (slow) x-ray film, then give a long exposure and look for any black patches. It is necessary to give an exposure of the order of 10,000 monitor units to be able to detect excessive leakage on radiotherapy film.

Head leakage is normally measured with the collimators closed to their minimum settings, as this will maximize the leakage readings. The plane perpendicular to the beam axis is assessed with the gantry at zero degrees, at a circular matrix of points. The greatest leakage is likely to be found on top of the gantry, due to backscattered photons. This region is most easily checked with the gantry positioned at 90°. Because of the relatively low dose rates being measured, a large volume ionization chamber (e.g., 30 cc) is best for integrated measurements, although a standard 0.6 cc Farmer chamber can be used.

For general assessment of radiation levels a survey meter can be placed in the room. Provided the readout display is large enough (something to consider when purchasing the meter in the first place), its reading can be checked using a camera system such as the one used to monitor the patient during treatment.

8.5.2 Treatment Accessories

Accessories (shielding block trays, wedge filters, devices for special techniques) are generally binary coded so that the accessory is recognized by the treatment machine when inserted. For patient treatments, these codes must match the codes generated for the intended treatment by the electronic record and verify system, or by the operator's manual selections. All supplied accessories should therefore be checked that they do generate the expected code when they are inserted in the treatment head, that is, that the linear accelerator uniquely identifies each one. Also, since the physical position of most of these accessories is critical for accurate treatment, the interlocking mechanisms should be checked to ensure that the accessory is seen as "in" only when it is correctly positioned.

8.5.3 Emergency Off Buttons

Emergency off buttons should be installed at several points within the bunker and also on the outside wall near the operator's console.

Some basic rules should be followed when determining the location of the emergency off buttons:

- They should not be located in the primary beam paths.
- A button should be easily accessible from either side of the treatment couch.
- Locations for buttons should afford a reasonably open view of the room.
- One button should be close to the exit of the bunker.

8.6 Patient Safety

As discussed in Chapter 1, it is not appropriate to define a dose limit for medical exposures where the patient is deriving a direct benefit from the exposure. However, the principle of optimization still applies—both the procedures followed and the equipment used must be designed to optimize the exposure. This optimization has two components, the delivery of the prescribed dose to the target and the minimization of dose to surrounding healthy tissues.

The radiation oncologist is responsible for prescribing a dose of radiation for which the benefits (tumor eradication or control) outweigh the risks (acute and delayed side effects due to damage to normal tissue, and possibility of induced cancer due to radiation exposure). The same applies to any imaging procedure used to assist in planning the radiotherapy treatment or for verification and correction of position during each treatment session—the benefit (greater precision in delivering the treatment) must outweigh the risk.

There are many steps involved following the decision to administer radiotherapy. These include

- Dose prescription (total dose and fractionation)
- Patient positioning and immobilization procedures
- Anatomical data acquisition (CT, MRI, or manual method)
- Definition of the target volume (encompassing the tumor) and dose-sensitive structures
- Determination of the appropriate treatment technique and beam arrangement
- Volumetric dose computation
- Plan evaluation and approval
- Fabrication of shielding shapes and treatment aids
- Transfer of data to the treatment machine (electronic or manual)
- Treatment verification and delivery

Unlike radiology and nuclear medicine procedures, radiotherapy treatments deliver very high doses of radiation, high enough to cause serious damage or even kill a patient if misadministered. Therefore, to ensure patient safety, each of these steps should be carried out according to established policies and procedures and must be supported by extensive quality assurance checks.

8.6.1 Treatment Planning

Treatment planning includes all but the first and last steps above, many of which are covered by software modules incorporated into a computerized treatment planning system (TPS).

Several reports, including references 12 and 13, have revealed that a very high proportion of serious radiation accidents in radiotherapy are the result of errors somewhere in the planning process. Whereas a treatment delivery error is more likely to occur on just one day or a few days before being detected, a planning error can affect the entire course of treatment. Planning errors are often systematic and subtle, and the most serious errors have resulted in large numbers of patients being mistreated.

This highlights the importance of quality assurance in treatment planning. References 14 and 15 provide comprehensive overviews of this subject.

8.6.2   Equipment Design

Radiotherapy equipment must meet strict specifications to ensure safety of patients and staff. The purchase and commissioning process for equipment such as a linear accelerator should require vendors to respond to specifications as outlined in a tender document, then demonstrate that the equipment meets those specifications before it is accepted by the user. The International Electrotechnical Commission (IEC) is the body responsible for setting the standards in all fields of electrotechnology. Regulators in most countries, as well as equipment vendors, observe the standards published in the relevant IEC documents.[16–18]

8.7   Miscellaneous Radiation Protection Issues

8.7.1   A Cobalt Source Becomes Stuck in the "On" Position

Cobalt-60 treatment unit manufacturers have each developed failsafe systems for ensuring that the radioactive source is returned to a safe storage position when treatment is completed, even if the primary mechanism for returning the source fails. For example, in a CGR Alcyon unit, the source is inserted on the circumference of a barrel, made of tungsten. Under normal operation, the barrel rotates to the "source out" position for treatment, then rotates back to a safe storage position when treatment is completed; if the normal source rotating mechanism fails, a recoil spring should bring the source back to its storage position.

Figure 8.9 illustrates a Theratron cobalt-60 treatment head; here, if the air pressure fails, the source should retract to the safe position shielded by the depleted uranium.

In the unlikely event that both the primary and backup methods should fail, the source will have to be manually returned to the safe position, following the method applicable to the particular treatment unit. The immediate and urgent concern, however, is to get the patient out of the path of the radiation and out of the treatment bunker. The best course of action is for two staff to enter the room. The first removes the patient from under the beam and assists the patient from the room as quickly as possible, while the second attempts to make the source safe. Once the patient has left the room, time is no longer critical. If there is any problem encountered making the source safe, leave the room and seek additional help. Under no circumstances use

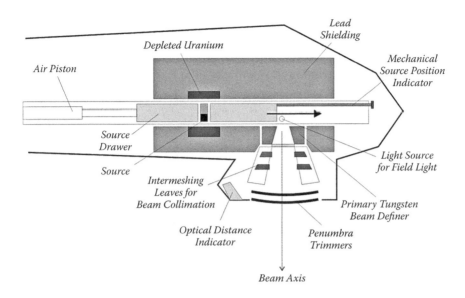

FIGURE 8.9
Treatment head of a Theratron cobalt-60 treatment unit.

the machine again until the failure has been investigated and the machine has been cleared for use.

The additional dose received by the patient will need to calculated and reported, based on an estimate of the additional time the patient was in the beam.

8.7.2 Activation of Materials Due to Neutrons from High-Energy X-Ray Units

For energies above about 10 MV, the production of photoneutrons also produces unstable (radioactive) isotopes in the treatment head of the linear accelerator, in particular the target and surrounds, collimators and shielding, and treatment accessories such as wedges or compensating filters.[19] Most of the materials become only slightly radioactive, and the half-lives of the products are quite short. The induced activity will be highest for any copper in the beam path, such as the thick copper backing of the target, due to the production of Cu-62 and Cu-64. However, because the induced radiation is collimated by the beam apertures, exposure levels remain low except for the region immediately beneath the treatment head. Even here, the exposure level is unlikely to exceed 0.2 mSv/h. If treating staff follow their usual work practices, they will not accumulate a significant exposure. A very conservative approach is to delay reentry to the bunker for some seconds or even a minute after treatment. This is not really warranted during normal clinical operation, but is justified after, say, a protracted period of high-energy beaming by physics personnel. Similarly, if service engineers need to work on or dismantle the treatment head, it is best if this is not done immediately after high-energy beaming.

8.7.3   Accidental Irradiation of Persons in a Bunker

If any person is inadvertently exposed to radiation, for example, if a treatment is commenced before it is realized that a staff member or patient's relative is still in the bunker, the machine should be switched off immediately. The circumstances of the irradiation (beam geometry, best estimate of the position of the person in the bunker, any other factors) should be recorded, to assist with an assessment of the dose received. The incident should be reported as soon as possible, according to the hospital's internal reporting policy, and also to the appropriate regulatory body if there is a possibility that the exposure level is above the threshold for external reporting. If the person is a designated radiation worker, the person's radiation monitor should be sent for immediate processing. The incident should be reviewed as to the root cause and contributing factors, and if necessary policies and procedures should be revised to minimize the risk of a repeat occurrence.

8.7.4   Shielding for Intensity-Modulated Radiotherapy (IMRT)

IMRT involves the use of much smaller field apertures, either as multiple static segments or in a dynamic mode where narrow field openings sweep across the target. In both cases, the total beam-on time is several times longer than for conventional therapy, and consequently, the proportion of secondary radiation (relative to primary) is increased. In addition, if high-energy (>10 MV) beams are used, neutron activation in the treatment head and also neutron dose to the patient will be correspondingly greater. The bunker design for a new facility will need to take this into account, if it is planned to use IMRT extensively. Existing facilities should review previous measurements and calculations to ensure the bunker is safe to be used for IMRT. The most pragmatic way to avoid protection issues in a bunker previously designed for high-energy beams is simply to use 6 MV for all IMRT treatments. With IMRT, the use of multiple modulated beams tends to offset the lack of penetrating power of lower-energy (6 MV) beams, so there is little or no clinical advantage in using higher energy.

8.8   Ongoing Quality Assurance

8.8.1   Safety Checks

Radiotherapy quality assurance checks fall into two categories:

1. Clinically oriented checks to ensure that the linear accelerator is delivering radiation within specifications, to maintain the accuracy and quality of patient treatments. These include checks of beam energy, output and uniformity, and mechanical alignment of gantry, collimator, and treatment couch.

2. Checks that relate to the safety of staff and the general public. In particular, "radiation on" lights, door interlocks, and last-person-out buttons should be checked at least weekly; emergency off buttons and manual methods of opening a neutron door or escaping via an alternative exit in the event of mechanical failure should also be checked regularly.

Each radiotherapy center should have a comprehensive quality assurance program in place, specifying the type of tests to be done, the frequency of testing, and the action to be taken when something is malfunctioning or out of tolerance. The results of quality assurance tests and subsequent action should be clearly documented and kept preferably for the life of the equipment but for a minimum of 5 years.

8.8.2   Radiation Surveys

It is not very likely that there will be any change to either x-ray machine or treatment room structure that will cause a change to exposure levels outside. One possibility is that lead shielding that surrounds the waveguide and treatment head is removed for a major repair and is not replaced correctly or is omitted afterwards. A complete check of head leakage could detect this, although a better strategy is to liaise closely with service engineers at the time of repair, then perhaps check a few appropriate points.

Shielding characteristics of a radiation room should not change over time, unless there are subsequent building works, for example, altering a maze to obviate the need for a neutron door. However, it is possible for cracks to develop in concrete walls if there is significant building movement; therefore, it is good practice to repeat a subset of the initial survey measurements annually. Points behind the primary barriers are usually the most important to check.

8.8.3   Staff Monitoring

All designated radiation workers in radiotherapy should be monitored with a personal radiation badge, comprising either a detecting element of film, TLD, or aluminum oxide—whichever is provided by the facility's chosen personnel monitoring service. However, it is to be expected that the majority of staff will receive little or no radiation exposure detectable above background levels during the course of their work, unlike nuclear medicine staff, for example, who inevitably accumulate low levels of radiation exposure in the course of their duties. In radiotherapy, the main purpose of the personal monitor is to provide a means of assessing exposure after a radiation incident, such as a staff member being left behind in a treatment room.

8.8.4 Staff Education

It is important that all staff, especially nonspecialist groups such as cleaners and tradespeople, are given basic radiation safety instructions, covering, for example, the difference between radioactivity and radiation from energized devices, and the purpose of the bunker gate. If their duties require them to enter the bunker, they should clearly understand when it is safe to do so.

8.8.5 Radiation Incident Reporting

No matter how much planning and care is taken, radiation incidents may occur from time to time. Radiotherapy treatments involve very high doses of radiation, often with a fine line between tumor control and major toxicity. Therefore, a misadministration of radiation can be very serious, potentially lethal. Most jurisdictions mandate that serious radiation incidents involving staff or patients be reported appropriately. However, it is important that each radiotherapy center also has an internal incident reporting policy and program, covering minor as well as major incidents, and including near misses—when a problem is detected and rectified in time to prevent an incident.

Regular review and discussion of these incident reports can reveal trends and weaknesses in current systems, thereby enabling procedures to be modified to lessen the chance of a repeat event.

References

1. International Commission on Radiological Protection, *1990 Recommendations of the International Commission on Radiological Protection*, ICRP Report 60. Oxford: Pergamon Press, 1991.
2. International Electrotechnical Commission, *Guidelines for Radiotherapy Treatment Room Design*, ed. 1.0 (IEC 61859). Geneva, Switzerland: IEC, 1997.
3. National Council on Radiation Protection and Measurements, *Structural Shielding Design and Evaluation for Megavoltage X and Gamma-Ray Radiotherapy Facilities*, NCRP Report 151. Bethesda, MD: NCRP, 2005.
4. McGinley, P. H., *Shielding Techniques for Radiation Oncology Facilities*. Madison, WI:, Medical Physics Publishing, 2002.
5. Martin, C. J. and Sutton, D. G. eds., *Practical Radiation Protection in Healthcare*. Cambridge: Oxford University Press, 2002.
6. Lo, Y. C., Albedos for 4-, 10-, and 18-MV bremsstrahlung x-ray beams on concrete, iron, and lead-normally incident, *Med. Phys.*, 19, 659, 1992.

7. Kersey, R. W., Estimation of neutron and gamma radiation doses in the entrance mazes of SL75-20 linear accelerator treatment rooms, *Medicamundi*, 24, 151, 1979.

8. McGinley, P. H., and Butker, E. K., Evaluation of neutron dose equivalent levels at the maze entrance of medical accelerator treatment rooms, *Med. Phys.*, 18, 279, 1991.

9. National Council on Radiation Protection and Measurements, *Structural Shielding Design for Medical X-ray Imaging Facilities*, NCRP Report 147. Bethesda, MD: NCRP, 2004.

10. Zacarias, A., Balog, J., and Mills, M., Radiation shielding design of a new tomotherapy facility, *Health Phys.*, 91, 289, 2006.

11. International Electrotechnical Commission, *Medical Electrical Equipment: Medical Electron Accelerators in the Range of 1 MeV to 50 MeV*, ed. 1.0 (IEC 60977). Geneva: IEC, 1989.

12. International Atomic Energy Agency, *Lessons Learnt from Accidental Exposures in Radiotherapy*, Safety Report Series 17. Vienna: IAEA, 2000.

13. International Commission on Radiological Protection, *Prevention of Accidents to Patients Undergoing Radiation Therapy*, ICRP Report 86. Oxford: Pergamon Press, 2001.

14. Fraass, B., Doppke, K., Hunt, M., KUtchger, G., Starkschall, G., Stern, R., and Van Dyk, J., American Association of Physics in Medicine, Radiation Therapy Committee Task Group 53: Quality assurance for clinical radiotherapy treatment planning, *Med. Phys.*, 25, 1773, 1998.

15. International Atomic Energy Agency, *Commissioning and Quality Assurance of Computerized Planning Systems for Radiation Treatment of Cancer*, Technical Report Series 430. Vienna, Austria: IAEA, 2004.

16. International Electrotechnical Commission, *Medical Electrical Equipment, Part 1, General Requirements for Basic Safety and Essential Performance*, ed. 3.0 (IEC 60601-1). Geneva, Switzerland: IEC, 2005.

17. International Electrotechnical Commission, *Medical Electrical Equipment: Medical Electron Accelerators, Functional Performance Characteristics*, ed. 2.0 (IEC 60976). Geneva, Switzernad: IEC, 2007.

18. International Electrotechnical Commission, *Radiotherapy Equipment: Coordinates, Movements and Scales*, ed. 1.1 (IEC 61217). Geneva, Switzerland: IEC, 2002.

19. *Clinac Technical Reference Guide*, P/N 1106795-02. Palo Alto, CA: Varian Medical Systems, 2005.

9

Radiation Protection in Brachytherapy

Ram Das

CONTENTS

9.1 Introduction

Brachytherapy is that mode of radiation therapy using small discrete radiation sources either in or on the patient. *Brachy* in Greek refers to short distance, and in brachytherapy the radiation sources are only a centimeter or so from the target within the patient, as opposed to teletherapy where the source is at a distance of about a meter. The sources are discrete in the sense that unlike in nuclear medicine, the sources do not metabolize and the integrity of the sources is maintained. They are encapsulated in materials that do not interact with body tissues or fluids. Sources with these features are referred to as sealed radioactive sources.

Soon after radium was discovered and isolated, brachytherapy with radium was the first radiation therapy modality to be used, as early as 1903.[1]

9.1.1 Radiation Sources Used in Brachytherapy

Although radium has been in use in brachytherapy for more than 75 years, the hazards with its use have long been recognized and substitutes have been sought. Ideally a brachytherapy source should have the following characteristics:

- The radioactive half-life should be long for temporary applications and short (a few days) for permanent implants.
- The physical form should be inert to body fluids and tissues.
- It should not have any gaseous radioactive daughters like radon is for radium.
- For the purposes of shielding and for dose reduction to tissues outside the tumor volume it should not emit high-energy photons.
- It should have no undesirable β emissions.
- It should be available in high specific activity so as to enable fabrication of higher-activity sources with dimensions of a millimeter or less.
- It should, preferably, be available in solid form.
- It should be cost effective.

In this context, apart from its long half-life, radium is no longer considered suitable as a brachytherapy source. Additionally, the buildup of helium pressure due to the radioactive decay (α particle) and heat sterilization has resulted in many accidents.[2-7] With the advent of nuclear reactors, many radionuclides can be made available with characteristics more suitable for clinical use. Most of the radionuclides currently used in brachytherapy and their characteristics are listed in Table 9.1.

9.1.2 Modes of Brachytherapy Delivery

When brachytherapy originated in about 1903, the needles and tubes of radium sources could be fabricated with only a few millicuries in each source. Resulting treatment doses were about 1,000 R (~10 Gy) in a day, with a dose rate of about 0.5 Gy/h to the treatment volume. Later, when it became possible to fabricate smaller sources with higher activity, the dose rate could be increased to about 1.5 Gy/min. Although vast experience was gained in about 75 years of brachytherapy with radium at a relatively low dose rate, better understanding of radiation biology has led to higher-dose-rate treatments that can now be practiced with different fractionation regimes. With a single miniature source of ^{192}Ir with an activity of about 370 GBq (10 Ci) the

TABLE 9.1

Physical Characteristics of Radionuclides Used in Brachytherapy

Radio-nuclide	Half-Life	Radiation Type	Energy (MeV)	TVLPb (mm), Concrete (cm)	Notes	Method of Production
^{226}Ra*	1,600 years	Photon	0.047 – 2.45 (0.83 avg.)	44, 23.3	Needles or tubes for interstitial and intracavitary therapy	Naturally occurring in decay chain of ^{238}U
^{222}Rn*	3.83 days	Photon	0.047 – 2.45 (0.83 avg.)	44, 23.3	Seeds for surface moulds, interstitial and intracavitary therapy	Naturally occurring Daughter of ^{226}Ra
^{60}Co	5.26 years	Photon	1.17, 1.33	40, 20.3	Needles, tubes, and beads for LDR and HDR therapy	Reactor irradiated ^{59}Co(n,γ)^{60}Co
^{137}Cs	30.0 years	Photon	0.662	22, 16.3	Needles, tubes, and beads for interstitial, intracavitary, and LDR therapy	Fission product chemically separated from spent fuel rods
^{192}Ir	74.2 days	Photon	0.136–1.06 (0.38 avg)	19, 13.5	Wires and seeds for interstitial and intracavitary LDR and HDR therapy	Reactor irradiated ^{191}Ir(n,γ)^{192}Ir
^{198}Au	2.7 days	Photon	0.412	11, 13.5	Seeds for surface moulds and interstitial therapy	Reactor irradiated ^{197}Au(n,γ)^{198}Au
^{125}I	60.2 days	Photon	0.028 avg.	0.1	Seeds for eye plaques and permanent interstitial implants	^{124}Xe(n,γ)^{125}Xe ^{125}Xe→^{125}I β
^{103}Pd	17.0 days	Photon	0.021 avg.	N/A	Seeds for permanent interstitial implants	^{102}Pd(n,γ)^{103}Pd

TABLE 9.1 (continued)

Physical Characteristics of Radionuclides Used in Brachytherapy

Radio-nuclide	Half-Life	Radiation Type	Energy (MeV)	TVLPb (mm), Concrete (cm)	Notes	Method of Production
^{90}Sr	28 years	Electron (β^-)	2.25 from ^{90}Y	N/A	Plaques for surface therapy, including eyes, beads for endovascular irradiations	Fission product
^{106}Ru	358 days	Electron (β^-) and photon (20%)	3.54 beta 0.512 gamma from ^{106}Rh	N/A	Eye plaques	Reactor irradiated ^{105}Ru(n,γ)^{106}Ru
^{32}P	14.3 days	Electron (β^-)	0 – 1.71 (0.69 avg.)	N/A	Wire for end ovascular irradiation	Reactor irradiated ^{31}P9n,γ)^{32}P
^{131}Cs	9.7 days	Photon	0.0295–0.0336	0.1	Seeds for permanent implants	Fission product

*Plus daughters.

treatment time can be shortened to a few minutes per fraction. By increasing the dose rate, fractionation is required to allow normal tissues to repair damage, something that would have occurred naturally in low-dose brachytherapy throughout the delivery. The different modes of brachytherapy and the radionuclides used are summarized in Table 9.2.

In the early days of brachytherapy, the encapsulated sources were inserted in tissues or cavities inside the tumor mass, directly or preloaded into applicators, as shown in Figure 9.1. Essentially the sources used had lower activities than in modern treatments. This method had many drawbacks. Radiation hazards included:

- The clinician could not use distance and shielding to reduce his or her exposure during application. He or she had to minimize the time of handling to reduce the exposure.
- Other staff, such as the anesthetist, nurses, technicians, and so forth, also received unnecessary exposure as no extra shielding could be provided.
- For radiographic verification and dosimetry, the patient with the radiation sources was required to be transported to other areas within the hospital.

TABLE 9.2

Different Modes of Brachytherapy[8]

	Dose Rate (Gy/h)	Typical Single-Source Activity	Isotopes in Use	Number of Fractions	Typical Applications	Comments
Low dose rate (LDR)	0.4–2	1500 MBq 10–15 MBq	^{131}Cs, ^{37}Cs ^{125}I, ^{103}Pd	2	Gynecological permanent implants	Common
Medium dose rate (MDR)	2–12	4.5 MBq	^{137}Cs, ^{60}Co	2	Gynecological	Not common
Pulsed dose rate (PDR)	0.4–2 in 1 h	37 GBq	^{192}Ir	Hourly, 3–40 h	Gynecological, prostate	Radiation safety hazard if unsupervised
High dose rate (HDR)	>12	400 GBq	^{192}Ir	2–5	Prostate, breast, gynecological	Most common

FIGURE 9.1
Manchester ovoids and uterine tandems using radium tubes.

- While the patient was in the ward, nursing staff and visitors received unnecessary radiation exposures.
- There was a risk of loss of radiation sources due to accidental dislocation or willful removal.

An additional drawback was the imperfect dosimetry resulting from the requirement to reduce radiation exposure by reducing time, leading to unsatisfactory source positioning, with corrections being well nigh impossible.

In practice there are nowadays only few applications where "life" radioactive sources are implanted into patients in the operating theatre. Seed implants with ^{125}I for prostate cancer are the most important example of this kind of application. The low-energy photons emitted by ^{125}I sources make shielding and radiation protection relatively easy.

However, in most brachytherapy applications life implants have been replaced by afterloading techniques, which are described in the next two sections.

9.1.3 Afterloading Systems

Henschke[9,10] first suggested the use of afterloading systems. This technique allows the insertion of hollow applicators or catheters (Figure 9.2) into the site of interest, followed by verification of the position and dosimetry to allow for realignment if necessary. A check radiograph used for planning is shown in Figure 9.3. When the patient is comfortably positioned in an appropriately shielded ward, the sources are preloaded in holders/pencils and manually inserted into the applicators. The introduction of this technique completely eliminated the radiation hazards in the theatre, radiology department, and to associated staff. However, the radiation exposure to staff in the ward, others associated with the source transfer, and visitors could not be reduced.

Remote afterloading systems completely eliminated the hazards of handling radiation sources for routine applications. The sources at the end of

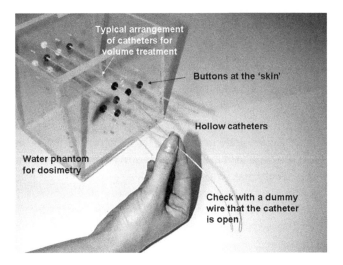

FIGURE 9.2
Test object with multiple hollow brachytherapy catheters. Radioactive wire of a predetermined length can be introduced into the catheters after they have been implanted into the patient.

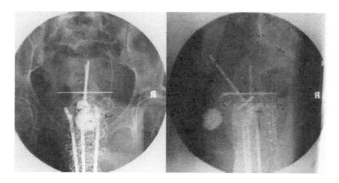

FIGURE 9.3
Radiographs of applicators in cervix.

drive cables are stored in shielded housing and can be driven in and out of applicators/catheters placed in or on the patient as preplanned. The source movements and the dwell time at any planned position within the applicator are remotely controlled by a microprocessor. In the manual and remote afterloading systems used for low-dose-rate (LDR) brachytherapy there is little scope for optimization of the dose distribution apart from altering the time for which a source train or source pencil dwells in a catheter. With high-dose-rate (HDR) brachytherapy the dwell position and the dwell time of the source in each catheter can be adjusted to obtain the desired isodose surface. Using this technique, HDR brachytherapy can be made more conformal than external beam radiotherapy.

9.1.4   Remote Afterloading Equipment: Low- and High-Dose-Rate Brachytherapy

9.1.4.1 Low-Dose-Rate Units

The majority of low-dose-rate remote afterloading units employ ^{137}Cs sources either as individual spherical sources or as "source trains" of various lengths with different source-spacer combinations. These units are mostly used for brachytherapy of gynecological tumors. The sources are stored in a shielded housing within the unit, and the required source configuration can be assembled and stored in intermediate storage. These are driven out into the patient catheters through the source transfer tubes connected between the unit and the patient catheters, either mechanically or pneumatically. The connection sockets of these catheters are such that only the correct transfer tube corresponding to the catheter number can be connected to each of them. If the connections become loose due to patient movement or other cause, the sources are automatically retracted to the intermediate storage. A backup battery in the unit provides the power to withdraw sources in case of failure of external power.

9.1.4.2 High-Dose-Rate Units

The high-dose-rate units are of two types. One type has three channels and is used exclusively for the treatment of cervix uteri. Many of these use ^{60}Co sources. The second type, which is used more prevalently, is a single-source unit using ^{192}Ir. A typical HDR unit and its components are shown in Figure 9.4. The mobile unit has the source storage, dummy source, and drive mechanisms on the upper section and the associated electronics in the lower

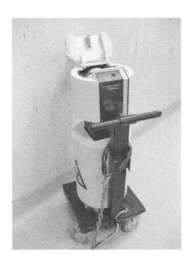

(a)

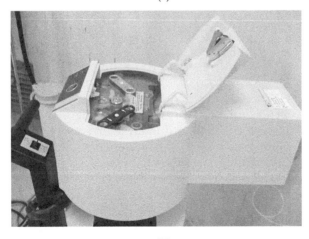

(b)

FIGURE 9.4
An HDR remote afterloading unit (*a*) contains source drives (*b*), numerous components (*c*), and a source assembly and drive cable (*d*). The components of (*c*) are (1) dummy cable drive, (2) source drive, (3) source position when not in use, (4) opto-pair for source control and detection, (5) Turret wheel controlling selection of catheter number, (6) indexer ring to connect and lock the transfer tubes, and (7) transfer tube connecting the catheter.

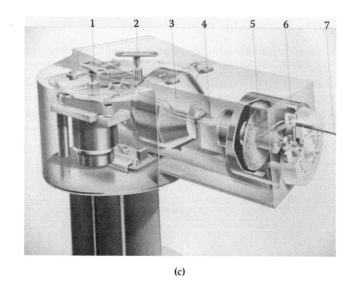

(c)

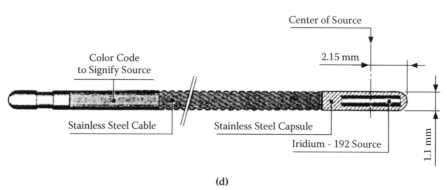

(d)

FIGURE 9.4 (continued)

section. Figure 9.4b shows the source drive, the dummy drive, and the source housing. More details of the system are shown in Figure 9.4c. The Pb shield housing the source and an identical dummy source are located at the center of the housing. The source assembly details and the dimensions are shown in Figure 9.4d. Active source has yellow color code. A sensing system (in this case an opto-pair) keeps control of source movement to and from the housing. The number of light pulses per second and the linear movement of the source cable are linked. Figure 9.4d shows the dimensions of the source and cable. The length through which the source moves is counted by the pulses, and during withdrawal if this number matches with the set drive while the source is driven out, the system accepts that the source is back in the housing. Some systems also incorporate an internal radiation monitor.

The turret wheel sequentially directs the dummy or source to the catheter channels on to the indexer ring with the catheter connection numbered sequentially. The transfer tubes shown in Figure 9.4c, through which the

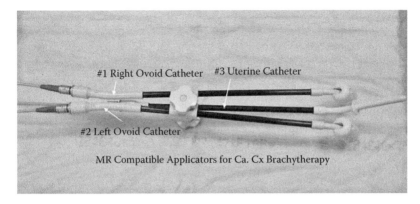

#1 Right Ovoid Catheter   #3 Uterine Catheter

#2 Left Ovoid Catheter

MR Compatible Applicators for Ca. Cx Brachytherapy

(a)

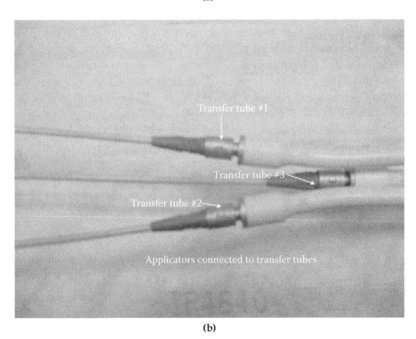

Transfer tube #1

Transfer tube #3

Transfer tube #2

Applicators connected to transfer tubes

(b)

FIGURE 9.5
Gynecological applicators connected to transfer tubes (*a*) and the details of the connection (*b*).

source is transferred from the unit to the catheters in the patient, are connected to appropriately numbered holes in the indexer ring. For gynecological treatments, the transfer tubes are numbered 1–3 and are to be connected to the right ovoid catheter, the left ovoid catheter, and the uterine catheter, respectively. The connecting ends are unique and the system will not allow incorrect connection (Figure 9.5). Once the treatment is initiated, the microprocessor of the control unit takes over the entire sequence of operation. From the control console, the source can be withdrawn by activating the treatment button or emergency button. There are many checks and controls in the system before, during, and after treatment. For example, if the dummy source check encounters any obstruction or difficulty in reaching the tip of the applicator, an error is indicated and the source will not be driven out.

The planned source positions and dwell times for each catheter can be transferred from the planning computer to the treatment unit either manually, on floppy disc, or via the local area network (LAN). During treatment, all dwell times will be automatically scaled as per the decay of the source.

9.1.5 Applications of Brachytherapy

9.1.5.1 Intracavitary Applications

The majority of intracavitary applications are for treatment of gynecological cancers of uterine cervix and vagina. An applicator is placed in the uterus and two applicators in the vaginal fornices (Figure 9.1). Traditionally radium tubes were placed in tandem in the uterine applicator, and hence it was known as uterine tandem. To shape the isodose surfaces to coincide with the surface of the vagina, the applicators in the fornices were shaped as ovoids, and hence the name vaginal ovoids. The resultant isodose surface was intended to cover the uterus and cervix, and the clinician positioned the applicator, or inserted needles into the treatment volume without the aid of imaging. Afterwards, the positioning was checked and treatment planning was performed using either two orthogonal radiographic images or, more recently, computed tomography (CT) or magnetic resonance (MR) images.[11] Using this method of retrospective assessment of the applicator position it was difficult to obtain a good geometry of the applicators and often the placement of applicators was far from satisfactory.

However, nowadays with image-guided brachytherapy and with the use of a single stepping source, much more acceptable dose distributions can be obtained with sparing of the bladder, rectum, and sigmoid. The applicators are placed at the required position with the guidance of imaging with an ultrasound probe (rectal or abdominal), with a C-arm x-ray unit, or with CT/MR scanners. The conventional LDR treatment is for two fractions, each about 72 h, separated by 7 days, giving a total of about 70 Gy, to a prescription point at which the uterine artery crosses the ureter and which is nominally situated 2 cm from the external os along the uterine canal and at 2 cm perpendicular to the canal (Figure 9.6). This point is known as Manchester Point A, following the group that proposed this system of dosimetry.[12] The prescription to points A (left and right) is a typical example of an applicator-based prescription where the dose is prescribed to positions that can be easily identified from the applicator seen on x-rays.

Although various HDR fractionation schedules are in vogue, most centers use five fractions of 6–7 Gy to Point A on different days, twice a week, or once in a week. However, there is now a trend that brachytherapy centers pre-

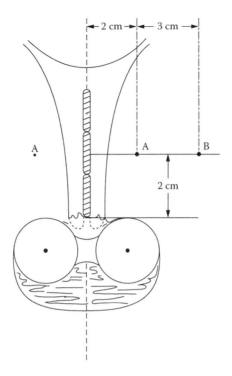

FIGURE 9.6
Definition of Manchester point A.

scribe a dose to the tumor surface identified by different imaging modalities. This is a very significant change from the older applicator-based prescription to an anatomy-based one.

There are a large number of different types of applicators available commercially,[13] including MRI-compatible applicators. Many gynecological cancers are treated with implants using steel or plastic catheters and templates. The dosimetry is based on geometry determined with MR, CT, or orthogonal x-ray films. Here again the dose is prescribed to the tumor surface. The multiple fractions are delivered either with one implant (once daily) or with multiple implants.

9.1.5.2 Interstitial Implants

For interstitial implants, discrete sources are implanted in the tumor volume as either a single plane, double plane, or volume implant. These are either temporary implants or permanent implants as done for prostate, breast, and so on (Figure 9.7).

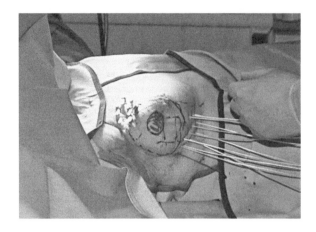

FIGURE 9.7
Temporary breast implant showing the afterloading catheters in place.

9.1.5.2.1 Temporary Implants

Generally long-lived radionuclides are used, such as ^{60}Co, ^{137}Cs, and ^{192}Ir (see Table 9.1). Hollow catheters are inserted into the tumor, generally under image guidance with ultrasound or fluoroscopy. The treatment is then planned with CT/MR/radiography images and an optimized plan is generated. Remote afterloading systems are widely used with HDR sources for the treatment. The sources are automatically positioned in the designated dwell positions for the designated times and returned to the storage.

9.1.5.2.2 Permanent Implants

The most commonly used radionuclides are ^{103}Pd and ^{125}I (see Table 9.1). A typical prostate implant may contain 80–100 seeds of ^{125}I or 120–150 seeds of ^{103}Pd. The seeds are implanted using image guidance with ultrasound or fluoroscopy. Transrectal ultrasound is particularly well suited. Mostly these are manual implants with preloaded needles. These needles are loaded with the radioactive seeds as preplanned based on ultrasound imaging. Remote afterloading systems are also available with on-line planning. In such systems the sources are also calibrated automatically. Generally, about 30 days after implant a postimplant dosimetry check is performed with CT images to determine the various dosimetry parameters, such as percentage volume of tumor covered by 100%, 90%, and so on of the prescribed dose (D_{100}, D_{90}); the percentage dose received by 100%, 90%, and so on of the target volume (V_{100}, V_{90}); and the doses received by critical organs.

9.1.5.3 Intraluminal Applications

Tumors that extend around annular cavities may be treated by a line source placed temporarily central to the cavity. The dose is uniform around the epithelium of the cavity, but decreases with depth. The main bronchus and the branches, esophagus, bile duct, and so on can be treated by line source applications.

9.1.5.4 Mould Applicators

Mould applicators offer an alternative to external beam radiotherapy with kilovoltage x-rays or megavoltage electrons when treating skin lesions. The main advantages are:

- Overall shorter treatment time
- Fast dose fall-off due to close proximity of the sources to the skin
- Easy coverage of complex concave and convex surfaces
- Ease to reach all body parts

Many superficial lesions are treated by a number of sources (dwell positions for HDR) placed over tissue-equivalent material over an area similar in size and shape to the target area but at a specified distance. The distance is arrived at on the basis of the depth of the intended target area under the skin. The mould is positioned appropriately, and for LDR treatments the mould is worn for a few hours per day as shown in Figure 9.8. HDR treatments can have a daily fractionation, that is, a few minutes per day.

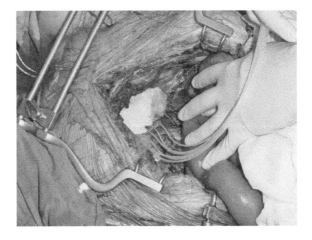

FIGURE 9.8
A surface mould in place for intraoperative treatment for rectal recurrence.

9.1.5.5 Intraoperative/Perioperative Brachytherapy

Patients can be treated by interstitial implants or mould applications intraoperatively with remote afterloaders. This is useful in cases where the organs are inaccessible externally. The tumor and tumor bed could be visualized and catheters or applicators are precisely placed in the target. The treatment could be a single fraction with the organs exposed or by multiple fractions with the incision closed but with the catheter leads outside the body. At the end of the treatment the catheters are removed.

9.1.5.6 Other Applications

Some brachytherapy treatments may be performed outside of a radiation oncology facility. In the treatment of uveal melanoma and other cancers of the eye, eye plaques with ^{90}Sr, ^{125}I, ^{106}Ru, or ^{103}Pd may be used in ophthalmology clinics (Figure 9.9). All procedures and safety measures described for LDR sources are applicable for these practices. However, in addition to this transport issues may arise if the sources/applicators are prepared in a different part of town.

Radiation sources have been used by many interventional cardiologists and vascular surgeons for prevention of restenosis; however, the procedure is losing favor. In these procedures a wire (^{32}P, ^{192}Ir) or a train of sources (^{90}Sr) is remotely inserted and positioned in the lumen of the artery over the desired length to deliver a prescription dose to the lumen walls. Generally this is performed after angioplasty, balloon dilation, or stent placement. In another technique a balloon is inserted and positioned in the lumen and liq-

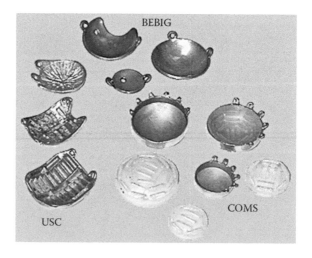

FIGURE 9.9
Examples of eye plaques.

uid ^{188}Re is injected into the balloon and withdrawn after the calculated time. For details of these techniques refer to Waksman.[14]

9.2 Potential Radiation Hazards in Brachytherapy

In all brachytherapy techniques there are radiation hazards to the patient, the hospital staff, and the public. The following section identifies these hazards and suggests appropriate measures for their reduction.

9.2.1 Patients

Patients are the direct beneficiaries of the treatment procedures, but exposure to other tissues at risk close to the tumor or organ of interest should be minimized. This is achieved by proper planning of the procedure, insertion of applicators with image guidance, postprocedure imaging, and optimized treatment dosimetry. Over the years, more than five hundred incidents have occurred in which patients received minor to fatal injuries.[15,16] These accidents can be attributed to human errors, mechanical failures, and lack of training and supervision. Examples of human errors include:

- Use of wrong source strength resulting in higher dose
- Wrong patient
- Wrong treatment site
- Wrong catheter connection

Since the use of incorrect source strength can result in serious over- or underdosages, the source strength of all brachytherapy sources should be verified at the user's facility. The calibration factors of the measuring system must be traceable to a national standard. It must be ensured that the correct source and source strength are used in the dose calculations. In the past, a number of units have been in use for representing the activity of the source—millicurie, K-factor, megabequerel, and so on. However, there are sources currently available with the same radionuclide but with different encapsulations and structure, resulting in sources containing the same amount of radioactive material, but having different radiation outputs. Hence, following the recommendations of the American Association of Physicists in Medicine (AAPM),[17] all source strengths are to be stated in reference air kerma strength (RAKR; in $\mu Gy \cdot hr^{-1} \cdot m^2$) along the equatorial axis of the source. Additionally, during commissioning and at regular intervals, all planning computers should be checked for the use of the appropriate source parameters specific to the type used.

There are reported incidences where a wrong person has been irradiated. Hence, it is essential to have appropriate management systems in place to verify and confirm the name and hospital number of the patient before commencement of the treatment.

In the case of multicatheter treatments, the sequential numbers of the catheters used for dosimetry should be maintained and the connections to the remote afterloading unit checked before commencement of the treatments. In optimized treatment planning with a single cycling source, the source dwell positions and dwell times will be different for the different catheters. Wrong catheter connections can result in serious and deleterious dose distribution errors. For example, in a prostate implant certain catheters in the periurethral region may have been planned with fewer dwell positions and shorter dwell times to reduce the urethral dose. If these catheters are connected to the treatment unit in the wrong sequence, the urethra may get an unacceptably high dose, resulting in urethral stricture.

In HDR units, the source positions in the catheter are dictated by the indexer lengths and dwell positions. Wrong entries of these values numerically or unit wise (centimeters instead of millimeters) can result in irradiating the normal tissues.

In the use of remote afterloading systems, mechanical failures and computer errors such as the following can lead to higher patient dose:

- Source detachment from drive cable
- Kink in applicator during source movement
- Loss of communication between control unit and treatment unit

These errors prevent the return of the source from the applicator and irradiate the patient until the source is detected and retracted.

9.2.1.1 Personnel Training and Supervision

Many incidents are the result of inadequate understanding of the procedure and the equipment by involved personnel. The radiation oncologist, the physicist, the radiation technologist (radiotherapist), and the nurse should be trained in the use of the afterloading units and safety procedures. A fatal incident in Indiana, Pennsylvania[18] was reported to be the result of lack of supervision by the radiation oncologist and physicist, both of whom were absent during the treatment. In this case, the patient is reported to have left the treatment room with a detached source in the catheter. Monitoring of the patient and the room at the end of the treatment would have identified the source still in the patient and steps could have been taken to recover it with minimum delay.

Other instances, with the use of preloaded radium applicators, include the patients pulling the applicator out and throwing it away, and patients being cremated with the sources still in vivo. To prevent any such incidences, all patients should be monitored immediately after the treatment. Similarly, in the case of permanent implants, the room, the linen (before being taken out of the room), and all other items should be thoroughly monitored after the patient has been discharged from the hospital.

9.2.2   Hazards to Staff

As mentioned earlier, the preloaded and manual afterloading applicators have resulted in unnecessary exposure to the staff. Since the use of preloaded applicators has been almost discontinued all over the world, this modality will not be discussed here.

Manual afterloading procedures, which are essentially LDR techniques, can still result in exposure to ward staff. As explained in Section 9.1.2, the dose rate of about 0.5 Gy·h^{-1} is used in intracavitary applications and the treatment time is about 72 h per fraction. In order to reduce the overall treatment time to a day or so and to treat more patients, many centers use higher-activity sources, giving about three times the Manchester dose rate. With use of ^{137}Cs pellets of about 30 mCi and between seven and ten pellets for intracavitary application, the dose rate at 1 m from the patient is about 1 mSv/h. A person spending an average of 30 min at this distance near the patient during a treatment will receive 0.5 mSv per patient. For an average of two patients per week, the total annual dose could be 50 mSv. If there are ten nursing staff sharing the workload, the average dose received by the individual could be 5 mSv per annum (25% of the maximum permissible dose). This is an avoidable dose with the use of remote afterloading systems. If prolonged nursing is required, it will be better to remove the sources and return them to the storage container, and reinsert them when it is safe to do so.

For remote afterloading systems, the treatment room housing the afterloading unit must be of a proper design so as to reduce the annual dose at other accessible areas to less than the permissible levels prescribed by the national or international bodies. Normally the interlocks at the door prevent any exposure to staff, as the sources will retract into the safe when the doors are opened or the interlock chains are removed. However, any failure in the retracting mechanism or kink in the applicators will prevent the source return. Retrieval of the source may result in unacceptable exposure to staff. The likelihood of this can be reduced by appropriate training and by following appropriate guidelines and procedures already laid down (see Section 9.3.9). Before commencement of the treatment it should be ensured that no person other than the patient is in the treatment room. Sample checklists for treatment planning and delivery are shown in Appendixes A and B.

A developing fetus is treated as a member of the public. Hence, any staff suspected or confirmed as pregnant should be treated as a member of the public, and the fetus should not receive more than 1 mSv. In brachytherapy, there is always a potential for high exposure, and therefore pregnant staff should be given duties where they are not exposed to radiation.

9.2.3   Exposure to the Public

Exposure to the public can be reduced to well below the permissible levels (1 mSv per year) by the appropriate design of the treatment room. However, visitors to the ward with manually loaded LDR brachytherapy patients could receive some radiation exposure, and so restrictions should be imposed on the number of personnel visiting and the period for which they can remain in the room (at a specified minimum distance), depending upon the measured dose rate. As a general precaution, children and pregnant women should be restricted from entering such wards.

In the case of patients discharged with moulds or permanent implants (such as ^{198}Au, ^{125}I, and ^{103}Pd), they should be instructed to keep children and pregnant women at greater distances or reduce the time spent with them. The distance and time should be specified based on the initial dose rate and the half-life of the radionuclide. Appropriate instructions should be given for the removal, storage, and safe return of sources in the mould applicators, and for retrieving any seed sources voided through urine (in the case of prostate implants). Patients should also be instructed to pick up any source or suspected items by forceps and not with fingers.

Usually there are guidelines issued by regulatory authorities for the activity of a specific isotope, which may still be in the patient when discharged from the hospital. This depends on the type of implant and the isotope used. The total air kerma strength (AKS) with which a patient can be discharged should be evaluated on the basis of the radionuclide (half-life), the social and living conditions, and the physical condition of the patient. For example, if the patient lives in smaller accommodations with a number of relatives (especially children or pregnant women), to avoid exposure to the larger group, he or she may not be discharged from the hospital. However, the exposure to family members may not be significantly high. In an affluent society, the calculated mean life-time dose for a spouse from the husband with about 1,500 MBq of ^{125}I in the prostate will be about 0.1 (range = 0.04–0.55) mSv.[19]

9.2.3.1 Burial or Cremation

Before burial, any removable source, as in mould applicators, should be removed from the patient. The burial of a permanent implant in the patient's body does not usually require any special precautions, because of the short half-life and lower activity of the radionuclide used for such implants.

During cremation the radioactivity could be released if the source integrity is lost in the environment of high temperatures. In such events, the staff of the crematorium could inhale the vaporized radionuclide (as with ^{125}I). For a typical prostate implant, about 1,500 MBq (40 mCi) of ^{125}I will be implanted. As little as 5 µCi (185 Bq) by inhalation or 2 µCi (75 Bq) by ingestion will result in an internal effective dose of 1 mSv.[19] Therefore, the recommendations[19] allow cremations only after 12 months of the implant. However, if cremation is required prior to this period, the prostate should be removed and stored in a steel container for the remainder of the 12 months and disposed of later.

9.2.3.2 Medical Emergencies

In some cases, the patient may develop complications requiring surgery. For example, after permanent implant of the prostate, the patient may develop urinary strictures requiring transurethral resection. If the operative procedure is essential, all the seeds must be removed, accounted for, and stored safely until they can be disposed of. In case of a prostatectomy performed 2 years after the implant, the risks are not relevant since the sources would by now have decayed to less than 0.4 MBq. Sometimes the patient may require another abdominal or pelvic surgery for other reasons, in which case the performing surgeon should be aware of the implanted seeds and should seek advice and clarification from the implant center. In order to facilitate the above, the patient should be given a card with the details of the implant that should be carried at all times for the first 2 years. A sample card is shown in Appendix C.

A suitably qualified medical physicist should be consulted in all these cases.

9.3   Radiation Safety Measures in Brachytherapy

9.3.1   Brachytherapy Facilities

Basic radiation safety in the use of brachytherapy sources starts with the design of the facility. For LDR techniques, the room housing the patient or, in an HDR facility, the room in which the HDR unit is housed for the treatment should have an appropriate design to reduce the annual dose to staff and public below the permissible levels at all accessible sites. The basic concepts of time, distance, and shielding are to be applied judiciously to calculate the annual doses. Achieving distance by housing brachytherapy units in large rooms can be difficult where space is at a premium, and hence reducing the occupancy in the vicinity and having adequate shielding are often the best ways to reduce the annual dose.

When selecting the location of the room, it is not advisable for it to be located close to an obstetrics or pediatric ward. Shielding should be considered not only for the walls, but also to the floor and ceiling, depending upon the utility and occupancy of the areas below and above the room, with due attention to load bearing of the structure. Any windows planned for a ground floor room need to have an area cordoned off to restrict access.

9.3.2 Low-Dose-Rate Brachytherapy Treatment Rooms

An LDR brachytherapy room will generally be part of a patient ward, and the layout should allow for the safety of the nursing staff and other staff in the ward and of the public visiting the ward. Further, the layout should provide for efficient nursing. In this context, the discussion will be limited to remote afterloading facilities. In addition to the general requirements for a patient-nursing room, there must be provision for the safe appropriate storage of the manual afterloading sources (if used), the afterloading unit, any source trains not incorporated in the unit, the source transfer or coupling tubes, and the emergency kit. An appropriate radiation detector should be installed in the room to detect any low activity source either in or outside the patient. Warning lights controlled by this detector should be displayed outside the room and at the nurse's station. The access door/chain to the room should have an interlock with the treatment unit. There must be provision for remote monitoring of the patient with CCTV and an intercom system.

9.3.2.1 LDR Brachytherapy Treatment Room Shielding Design

For a room of given dimensions, the average effective dose (mSv) per year can be calculated based on the number of treatments per year (number of patients multiplied by the average number of fractions per patient), the average activity used per treatment, the average treatment time, the occupancy factor (fraction of the day the area is occupied), and the distance from the patient bed to the point of interest. If this calculated value is higher than the design limit (dose constraint), shielding of the wall must be increased with appropriate material to reduce it below the dose constraint D_{cons}. This general approach of determining workload and occupancy is similar to the concepts that have been discussed in the context of external beam radiotherapy in Chapter 8.

The average dose, D_{av} (mSv), to a person at an accessible location outside the treatment room, before attenuation through the wall, can be estimated by the equation

$$D_{av} = W \times U \times T \tag{9.1}$$

where U is the use factor, T is the occupancy factor, and W is the total workload (mGy) per year given by $W = W_1 + W_2$ (W_1 being the contribution from the primary radiation and W_2 the contribution of scattered radiation reaching the location from proposed walls). The use factor is the fraction of the treatment time during which the radiation is directed toward the wall, and in brachytherapy this will be one for both the primary radiation and the scattered radiation (also see Chapter 5 and Appendix A).

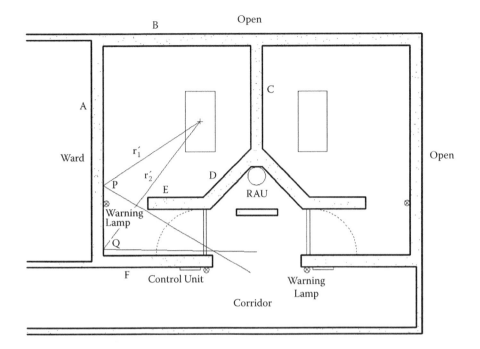

FIGURE 9.10
A sample layout for an LDR brachytherapy treatment facility with two patients.

Example Calculation

The following sample shows the calculation of shielding requirements for an LDR brachytherapy room with a sample layout for a remote after-loading system, for treatments of two patients, as shown in Figure 9.10.

Primary Barrier

$$W_1 = N \times T_x \times S_{av} \times \text{RAKR}/d^2 \qquad (9.2)$$

where N is the total number of patients treated in a year, T_x is the average treatment time per patient (h), S_{av} is the average number of sources used per patient, RAKR is the mid-year strength of the single source (μGy·h^{-1}·m^2) and d is the shortest distance (m) to the wall from the average source position during treatment.

As an example, let us assume that $N = 50$, $T_x = 40$ h, $S_{av} = 10$, and RAKR $= 84.82 \, \mu$Gy·h^{-1}·m^2 for the Cs-137 pellet used in the afterloading system.

		Barrier Thickness				
Location at	**Distance (m)**	**W1 (mGy/year)**	**T**	**Estimated Annual Dose (mSv)**	**TVT s[a]**	**Thickness of Concrete (cm)**
Wall A	3.5	138.5	1	138.5	2.14	35
Wall B	2.5	271.4	1/16	17	1.23[b]	20
Wall C	2.5 (patient in next room)	271.4	1	271.4	2.43	40
Wall D	2	424	1/4	106	2.02	33
Floor	5 (waist level at room below)	67.8	1	67.8	1.83	30
Wall E + F	5	67.8	1/4	17	1.23	20

[a] TVT s: Required to reduce the dose to D_{cons} taken as 1 mSv (see Table 9.1)·

[b] If the annual D_{cons} for the public is taken as 0.3 mSv, then this factor will be 1.78, resulting in a wall thickness of 28 cm concrete.

Secondary Barrier

In the example shown in Figure 9.10, scattered radiation to wall F and at the door only is to be considered. The first scatter reaching the wall F and the door is only from wall A. The scatter dose (D_{sc}) from point P, from an area of 1 m x height of the room (3.5 m) at the door, will be given by

$$D_{sc} = W_{1(r1)} \times \alpha A1/(r_1')^2 \tag{9.3}$$

where $W_{1(r1)}$ is the annual dose at P distant r_1 from the source, α is the reflection coefficient for scatter,[20] A is the area of the wall at distance r_1 (1 m wide), and r_1' is the distance from P to the door. The calculation could be repeated every ten or twenty degrees and summated to obtain the total scattered dose. In the example, the scattered dose at the door will be less than 1 mSv. At wall F the dose may be about 6 mSv, requiring 2.6 HVT of concrete for the scattered radiation to be reduced to less than 1 mSv. The HVTs for the scattered radiation can be obtained from NCRP 49.[20] For the primary barrier a thickness for wall E + F as 20 cm was obtained, of which 15 cm could be assigned to wall E. The scatter component may require 10 cm for wall F, which is more than the remaining 5 cm for the primary shielding. Hence, the design may have 15 cm for wall E and 10 cm for wall F.

Note: The advantage of a room at the top floor is that the room could be provided with windows without compromising on radiation safety. If it is desired to modify existing room walls, the floor and ceiling have to have additional shielding as calculated. As a space-saving measure, steel plates or lead sheets of appropriate thickness could be in-built into the walls.

Mobile Shields

In some cases where it may not be possible to provide the required shielding to the walls and door (e.g., for cost, structural, or space considerations) mobile shields can be fabricated and used at the patient's bedside. These shields should be at least 30 cm above floor and with a height of 150 cm and are to be placed as close to the bed as possible. Being close to the source of radiation makes it possible to use an overall smaller area of shielding material and as such less weight. The shields are generally made of lead sandwiched between steel plates in a steel frame with appropriate castor wheels, which could be locked on to the floor. The thickness of lead required is calculated as discussed earlier. They should be made in sizes that can be easily moved about.

It is advisable not to have more than one patient in brachytherapy rooms at the same time.

9.3.3 High-Dose-Rate Brachytherapy Treatment Rooms

For many reasons the most ideal location for an HDR unit is in an operating room with imaging facilities. First, moving patients around subsequent to positioning/insertion of applicators or catheters can result in movement of applicators within the cavity, especially for intracavitary and intraluminal applications. Hence, the geometry of the applicators/catheters may significantly change from intended and planned (from imaging), leading to the wrong treatment. Second, any envisaged intraoperative brachytherapy (IOBT) could be performed only if the HDR unit is in an operating theatre. An integrated brachytherapy suite with the operating table that can be rolled into a CT/MR gantry for imaging and can be used for HDR treatment in the same room will be ideal.

A sample layout of an HDR treatment room is shown in Figure 9.11. The design with a maze avoids the use of heavy shielding on the door. It is a compromise between the requirement for a heavy door that would require motorized movement and the distance persons have to walk/run to attend to the patient in an emergency. The room should be large enough to accommodate all accessories used in the theatre and should be able to accommodate the large number of surgical and allied staff involved in the procedure. Of particular importance is the availability of fluoroscopic imaging (C-arm). The corridors should be wide enough to transport a patient trolley/bed with ease. The shielding thicknesses of the walls could be calculated the same way as for the LDR unit.

Example Calculation

Workload Estimation

Number of patients per year treated for site 1 N_1
Number of average catheter per site 1 C_1
Average number of fractions per site F_1
Average treatment time per fraction (min) t_1
Average source strength ($cGy \cdot h^{-1} \cdot m^2$) Λ
Distance to point of interest P (m) d

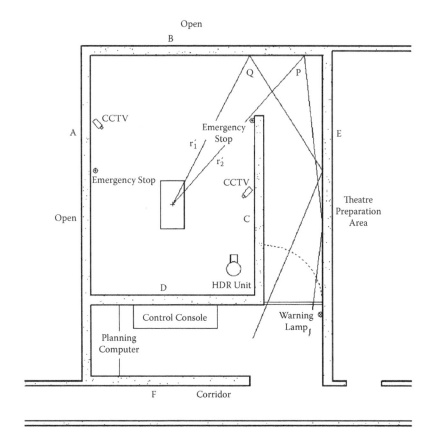

FIGURE 9.11
A sample layout for an HDR treatment facility.

The annual effective dose at P due to the primary may consist of three components. The first component from the primary when the source is in treatment position is given by

$$D_{PP} = N_1 \times F_1 \times t_1 \times \Lambda/(100 \times 60 \times d^2) \text{ Sv} \tag{9.4}$$

For HDR source such as ^{192}Ir, with source changed every three months, the average source strength could be calculated as for around day 45 and t_1 adjusted for this strength. The treatment time does not include the transit time to and from each catheter. This also should be accounted for. The second component of dose due to transit (not within the catheter for different dwell positions) can be estimated as

$$D_{P(t)} = N_1 \times C_1 \times F_1 \times 2 \; tr_1 \times \Lambda/(100 \times 3600 \times d^2) \tag{9.5}$$

where tr_1 is the time (s) taken for the source travel from the parking position to reach the first treatment position in a catheter. For most systems this time will be 3–5 s. The third component, D_{Pq}, is the dose due to all quality assurance tests and calibration procedures. This depends upon the type of tests each center carries out regularly.

The total annual dose from the primary contribution from treatment site 1 to n is given by

$$Dp = \sum_1^n \left[N_1.F_1.t_1.\Lambda / 100 \times 60 \times d^2 \right] + \sum_1^n \left[N_1.C_1.F_1 2t_{r1}\Lambda / 100 \times 3600 \times d^2 \right]$$
(9.6)

The scatter component could be calculated the same way as with equation (9.3). The required shielding thicknesses then could be calculated as for the LDR room with appropriate factors to reduce the annual dose to D_{cons}. The thickness required for the floor or ceiling could also be calculated in the same way.

Some local regulations may have D_{cons} as a dose rate in μSv·h⁻¹. In such cases the average dose rate for a treatment, at the distance of interest for primary/scattered, should be taken into account in arriving at the shielding thickness rather than the annual dose.

The long maze will ensure that only second scatter reaches the door, and it will not be necessary to have a heavy door. A physical barrier with an interlock will suffice. The control console, treatment planning system, and so on could be accommodated in the adjacent room. The treatment room should have emergency stop buttons conveniently located inside the room. There must be provisions for monitoring the patient and associated anesthesia monitors and the HDR unit as well as intercom system for communication with the patient. Even if the HDR unit has an in-built radiation monitor to indicate the status of the source location, it will be better to have an independent monitor in the room with a warning light displayed at the door.

9.3.4 Other Sources

Though many facilities use afterloading systems, manual or remote, there may be some still using individual sources for moulds, eye plaques, and so forth. Such users should keep all sources in a shielded safe in a specially designated room. The room needs to be lockable and a source register must be kept. If source preparation is required, it is good practice to not perform the task alone. In addition to the possibility of an independent second check, this practice avoids hazards of a single person fainting in an isolated room with radioactive sources. In this respect, source storage and preparation areas should be considered similar to confined spaces in occupational health. If working in pairs is not possible, it is essential to let someone else know that a person is working with radioactive material.

The different types of sources should be stored in easily identifiable compartments in the safe or individual containers. Lead (5% antimony) blocks may be used with appropriately drilled holes to hold any needles/tubes and the inactive eye of the source projecting above to facilitate easy pick up of the sources. All such containers/blocks should be appropriately labeled. It is essential to keep an inventory of all sources and a record of the movement of the sources from storage, to preparation, to the patient, and return to safe. After use all sources should be identified and physically checked for any damage. Color coding of sources may be an easy way to identify them. When not in use, all sources must be locked away securely. Proper entry should be made of any source returned for disposal or disposed as waste. Handling of all sources should be performed behind an L-bench with sufficient shielding and a lead glass viewing window. The sterilization of sources should be performed only as per the written procedures given by the manufacturers.

The location of the storage and handling area should be close to the facility in which the sources will be used to minimize the distance of transport. This room should have good ventilation and an exhaust fan that should be switched on at least a half hour prior to the use of the room. During transport all containers must have appropriate labels and should be locked. Under no circumstances should they be left unattended in corridors or places accessible to the public. Particular consideration must be given to the use of public lifts (stuck in a lift with a radioactive source and members of the public is a potentially dangerous situation)—as such, it is preferable if treatment and storage areas are on the same floor. An extra empty storage container and long forceps must be available in the ward or application area, to pick up and store any displaced source. For further reading see Parry et al.[21]

Finally, it is important that emergency services and fire brigades have a plan of the storage areas for radioactive materials. Regular site visits by relevant officers can facilitate this.

9.3.5   Treatment Planning

Treatment planning comprises not only the dose calculation and optimization, but also the entire process of selection of the patient, patient preparation, the source/catheter implantation (image guided or otherwise), imaging for dosimetry, and treatment with appropriate source(s). In terms of radiation protection, good treatment planning fulfills two roles: it ensures the correct dose is given to the target while minimizing the dose to normal structures, and it reduces the likelihood of accidents. The institution protocols should be strictly followed in the selection of the patient. Otherwise, the patient will be unnecessarily irradiated or given the wrong dose.

9.3.5.1 Patient Preparation

The patient must be prepared according to the requirements for specific anesthesia. For example, a patient on aspirin dosage may not be suitable for general anesthesia and has to be off the medication before the implant can be done. Similarly, if the patient had no bowel preparation, it will be futile to attempt a rectal ultrasound-guided prostate implant because of poor imaging or the implant will be of very poor quality.

The patient should be positioned in such a way that imaging and insertion of the applicators pose no hindrance. For example, for treatment of prostate, the patient is placed in the lithotomy position with legs in the 90°- 90° position and the rectum almost parallel to the operating table top. This helps to get a better image of the prostate with rectal ultrasound and to lessen the interference of the pubic arch during needle implants.

9.3.5.2 Catheter Placement

Prior to placement, appropriate steps should be in place to identify and check the patient's name, site of application, and so on. If working with a preplanned catheter placement, as for ^{125}I implant for prostate, the patient should be positioned the same way as that done for preimplant imaging. Image-guided placement should achieve more accurate results and improved dose distribution. Many computer programs can generate optimized plans, but no computer can compensate for a poorly placed implant. When multiple fractions are planned, all catheters must be anchored to a device attached to the body/ treatment area such as templates and buttons. For intraluminal applications, it is better to use catheters that have provisions to anchor the catheter to the wall of the lumen or cavity. It should be ensured that enough of the catheters project outside the skin, to make connections to the afterloading unit or for easy introduction of the source trains. Except during imaging and treatment all catheters should be capped to avoid any blockage or contamination by foreign material.

9.3.5.3 Imaging

All imaging for dosimetry should be done with the patient in the same position as for treatment. The personnel conducting the dose planning may not be the people executing the treatment. There should not be any ambiguity or errors in the order in which the catheters are numbered for planning and for treatment. Therefore, there must be an accepted norm to number the catheters. For example, for prostate implants, the numbering might start from the most anterior catheter on the patient's right and sequentially to the left in the same row. Then continue with the next row from the patient's right, the last number being the catheter on the posterior row most lateral on the left. Lack of such norms has resulted in wrong catheter connections to the HDR unit.[17] Appropriate x-ray, CT, or MR markers should be placed correctly in the catheters to identify the extent to which the catheter is positioned in the organ of interest.

9.3.5.4 Dosimetry

It should be ensured that in the planning software the correct menu is selected for dosimetry and the patient's demography data are correctly entered. The operator should check that the image file/radiograph selected is for the correct patient. In the case of radiographs, the magnification factor should be checked and verified. Selection of source configuration data and other parameters must be independently verified by another person. All catheters should be identified sequentially and source positions decided based on organ delineation and cover. In the case of HDR/PDR applications, the indexer length used for the catheters should be verified and input. The prescription sheet should be checked before inputting the dose points and the prescription dose. All the above should be checked independently by another qualified person. Once a satisfactory plan is generated and approved by the radiation oncologist (RO), the plan and relevant data should be printed out and the RO should check and sign. This plan data should be used to set the remote afterloading machine for treatment or to load sources for manual systems. A sample of a checklist is given in Appendix A.

9.3.5.5 Treatment

Before placing the sources in the catheters (manual loading) or connecting the catheters for the remote afterloading systems, the correctness of the patient and the site of treatment must be ensured to avoid accidental irradiation of a patient/organ.[17] All catheters should be connected sequentially to the unit and locked properly and checked by a second qualified person. Run a dummy/check source to ensure that all catheters are patent. The pre-treatment record or set parameters should be printed and compared with the treatment plan with due scaling for activity correction for decay of the source(s). All interlocks and shielding should be checked to see that they are in place and that no person other than the patient is in the treatment room. Operators should check that the CCTV system is properly focused on the patient/HDR unit, and so on. The presence of a qualified radiation oncologist (and an anesthetist if the patient is anesthetized) in the control room must be ensured to assist with handling emergencies. Some direct-reading radiation monitors should be readily available before starting the treatment. A sample checklist is given in Appendix A.

9.3.6   Written Procedures

Protocols for various treatment modalities, with additional dose to any procedure justified, must have the approval of the hospital ethics committee. These must be readily accessible to all involved personnel. From a radiation safety point of view, dose prescriptions, dose distribution and coverage of target volume, dose limitations to organs of interest, and so on must be strictly followed. The prescription provides the justification for administration of radiation, and all other steps constitute the optimization process required by radiation protection regulations.

9.3.6.1 Imaging and Planning Procedures

There must be written guidelines for imaging procedures such as for radiography, CT/MR, ultrasound, and so on. For example, the focus-film-distance (FFD), focus-skin distance (FSD), and various kV mA combinations used in radiographic procedures for better visualization of applicators and x-ray markers should be addressed in these guidelines. Similarly, for CT/MR imaging, the slice thickness and spacing, margins to be covered, type of markers to be used with the applicators, and so forth should be included in the guidelines. These checks will avoid errors in reconstruction of the images in the planning computer.

9.3.6.2 Treatment Planning

In many hospitals, the treatment planning might be performed by different physicists, dosimetrists, or radiotherapists at different times. Hence there must be written procedures for planning for the different types of applications practiced, and all planners should be familiar with the procedures. Procedures for checking the plans should be in place. An example of a form for checking the plans is given in Appendix A. A second planner must check all parameters.

9.3.7   Treatment Procedures

Treatment procedures and responsibilities must be clearly written and must be available at the treatment console. In the case of LDR treatments, generally treatments are initiated or interrupted by the charge nurse in the ward. The procedures should clearly identify and illustrate the various responsibilities. All the catheter connections should be checked and their positions in relation to body marks (in the theatre) should be verified and recorded. Treatments should not be initiated if the applicator positions are moved by more than the stipulated distances.

In the case of patients with ^{125}I or ^{103}Pd seeds in the prostate, instructions must be clear regarding monitoring of urine. There must be clear instructions on how to identify and handle any source dislodged from the patient or from the LDR remote afterloading unit. The availability and location of handling and storage equipment and radiation monitors must be clearly displayed. Appropriate instructions are to be posted with the details of the person(s) to be contacted in a radiation emergency.

Responsibilities must be clearly identified regarding removal of patient from HDR unit or discharge of patients with implants. The amount of activity with which a patient can be discharged (^{198}Au, ^{125}I, ^{103}Pd, etc.) must be established and included in the procedure documents. All patients with other removable sources and personal effects should be monitored before leaving the treatment room. It is obvious that any radiation warning signals must not be ignored, but the presence of a source should be confirmed with another monitor. Ignoring the signals could lead to fatal accidents.[18] All moulds and applicators should be thoroughly monitored, and once the sources are dismantled from them, they should be properly identified and stored back in the right places.

9.3.8 Redundancy Checks

All systems used in brachytherapy should have redundant checks. From patient identification and verification until the patient is discharged, redundant checks will be a safe practice. All treatment parameters used and the generated plans should be checked independently by another qualified person and approved. In the case of LDR treatments, the type and activities of the source should be checked and verified. For remote afterloading units, the parameters set on the unit must be again checked and verified with the generated plan. All the catheter connections have to be verified independently by another person.

There should be a radiation warning system installed in the treatment room/control console to indicate the status of the radiation source(s) in the treatment room. However, an independent radiation monitor should be used to ascertain that the source is back in the safe storage and not in either the patient or any of the transit connections.

9.3.9 Emergency Procedures

Based on the type of source(s) and the devices used, the application, and the location of the treatment room, a written emergency plan must be prepared and personnel involved in brachytherapy should be familiar with the procedures. The procedures should be posted at the control unit, nurses' station, and other relevant areas. This plan should integrate with other emergency procedures in the hospital (such as fire emergency).

All staff involved should have training with mock emergency situations, including the use of radiation monitors and handling equipment. Training exercises must be arranged at regular intervals with provision to train new personnel. In the case of HDR units, it may be better to have the exercises arranged during source exchanges, when all checks could be performed with the dummy source. Records of the training, including list of participants, must be maintained.

A sample sequence for an emergency retrieval of an HDR source is given in Appendix D.

9.4   Quality Assurance

Quality assurance (QA) provides an indispensable and invaluable role in preventing patient over- or underexposure and avoidable radiation exposure to staff and public and in minimizing downtime of the equipment used. Routine QA checks may detect any anomalies in the equipment, which can then be remedied. Although regulations stipulate certain QA programs, it is advisable to customize the program to the local scenario and clinical practice. QA begins with setting goals and tolerances for dose delivery. For example, the American Association of Physicists in Medicine recommends the following tolerances:[22]

- Source positional accuracy ±2 mm in relation to the applicators
- Source strength calibration accuracy ±3%
- Dose calculation accuracy ±2%
- Dose delivery accuracy 5%–10%

The following section discusses some QA procedures relevant to brachytherapy equipment.

9.4.1   Low-Dose-Rate Brachytherapy

9.4.1.1 Acceptance Tests
Any new equipment and sources should be tested to check that they are within the specifications advised by the manufacturer. These checks include the storage, intermediate storage, source drive mechanism, source positioning in the catheter, catheter connection and interlocks, treatment timer accuracy, emergency source withdrawal, emergency battery status, source calibration, and source leak tests. A sample checklist is shown in Appendix E.

9.4.1.2 Inventory

In the case of manual afterloading systems, a proper inventory of all sources ordered, received, used, and disposed of must be maintained. In cases where sources are ordered based on a preplan, as for prostate implant with ^{125}I seeds, the number of sources used in the plan and their activities must be checked with the order for the sources and ensure the correctness of the order. If preloaded source trains are ordered, each train must be checked before ordering as well as after receipt. An autoradiograph of each source train/needle should be taken, verified, and filed with appropriate patient details.

9.4.2   HDR Brachytherapy

9.4.2.1 Acceptance Tests

The required QA checks at the time of installation include all electrical systems and controls. Perform a radiological survey of the installation to ensure the adequacy of the shielding. The catheter connections should be checked for the various types of applicators to be used, their interlocks, indexer lengths (the maximum source-out distance in a catheter/applicator), source position accuracy, timer accuracy, door interlocks, patient communication system including CCTV, and autoradiographs (if possible, superimposed on catheter radiographs) of the source position within an applicator. A sample checklist is shown in Appendix E.

9.4.2.2 Regular QA Checks

Every day before the commencement of treatment all doors and functionality of all the radiation monitors must be checked. The periodicity and types of QA tests to be performed must be established and followed routinely.

9.4.2.3 Source Strength

The serial number of the source must be physically checked and verified with the calibration certificate. All new sources must be calibrated with a dosimetry system with calibration factors traceable to a national standard. The calibration must also be checked independently by another physicist. If the measured value is outside 5% of the stated value in the calibration certificate, the discrepancy should be investigated and discussed with the manufacturer before the source is accepted. A detailed description of these QA tests is available in AAPM TG 43.[22]

9.4.2.4 Applicators and Catheters

All reusable applicators and catheters must be checked at regular intervals for any crack or wear and tear. Any screw connections must also be checked for loss of screw thread, tight fitting, and so on. Patency of all applicators/catheters must be established before insertion, to prevent the source from becoming stuck in the applicator.

9.4.3   Planning Computers

At the time of commissioning, the algorithms and source parameters used in the calculations must be checked. Generally, all vendors follow the recommendations given in TG 43[22] and its update.[23] The correctness of the parameters such as radioactive half-life, radial dose function and anisotropy geometry factors must be verified. If any new type of source is used appropriate factors must be input.

Spatial accuracies in reconstruction of the images from radiographs/CT/ MR/ultrasound must be ascertained. Periodic calibration of any digital image input device must be performed.

Using a standard plan with a single catheter and a couple of source positions, the dose values generated by the planning computer for a few points must be verified with a manual calculation using the same source parameters. This could be performed at the time of commissioning and annually.

Some additional comments on quality assurance are given in Chapter 5 of this book. Readers are also referred to Thomadsen[24] for a detailed discussion of planning system QA in brachytherapy.

9.4.4   Imaging Systems

The QA program for imaging systems should follow those procedures outlined in Chapter 6.

9.5   Transport of Radioactive Substances

9.5.1   Transport to and from the Hospital

All radioactive substances transported within the country and outside must follow national and international regulations. For general guidelines refer to IAEA TS-R-1.[25] Local regulations are likely to exist and must be followed. The transport of radioactive material comprises of all operations and conditions associated with and involved in the movement of radioactive materials: these include the design, manufacture, and repair of packaging, and the preparation, consigning, loading, carriage including transit storage, and unloading at the final destination of the package. However, for the purpose of this section, the discussion will be restricted to receipt and dispatch of the transport container.

On receipt of a radioactive consignment, the physicist or the radiation safety officer (RSO) must verify the addresses of the sender, consignee, and the declared contents and ensure that the package is correctly delivered. Check for any damage to the container. In case of any damage, contact the supplier and ensure the integrity of the source container. Test the container for any contamination. Measure the maximum surface dose and location and ensure the value is within permissible limits. Store the container under lock and key until the sources are used.

Generally, sources with long half-lives are returned to the supplier after use. This should be a condition of purchase. In most brachytherapy facilities these sources will be ^{192}Ir, ^{137}Cs, ^{106}Ru, ^{90}Sr, and infrequently ^{125}I. The shielded containers and the drums/packages in which the sources arrived can be used for the return. The sources must be placed in the container as per the instructions from the supplier. Improper source placement can lead to unsafe transport.[15] Sources with low activity must be tested for any contamination before the placement.

The source container must be secured properly as per its design, placed in the outer container/drum/package and sealed as instructed. All the necessary paperwork and appropriate labeling of the transport container must be completed. The maximum allowable radioactivity in each package of the various nuclides is listed in Section IV of TS-R-1.[25] Radionuclides are also classified according to chemical form, toxicity, fissile/nonfissile, and so on as given in Section V of TS-R-1.[25] For example, the UN classification number for ^{192}Ir is 2915. The types of packages are also classified[24] as type A, type B, and type C depending upon the type and quantity of radionuclide. In brief, type A containers are designed to withstand minor accidents while type B containers must be fireproof for an extended period of time and be able to withstand significant impact. If a 60-cobalt source for radiotherapy is transported it would be shipped in a type B container.

Another classification is for the labels[25] white I, yellow II, yellow III, and white fissile. This also is based on the transport index (TI), which is defined as the maximum dose rate at 1 m distance from the surface of the container in millisieverts per hour multiplied by 100. Together with the maximum dose rate at the surface of the package the transport index determines the type of label as shown in Table 9.3. For ^{125}I seeds white I must be used, whereas for ^{192}Ir (for the types of sources used in a hospital) yellow III must be used. These labels must be properly completed and be pasted on either side of the container/drum without masking any other markings.

Sources must be transported only by appropriately licensed personnel. The container should be handed over only to an authorized transporting agent, after ensuring his or her photo identity. In one instance (personal communication), a transport package with ^{192}Ir (exchanged source from an HDR unit) was collected by an agent collecting nonradioactive waste and the package was disposed of in the municipal rubbish dump (on recovery, only minor damage to the external drum was noticed).

TABLE 9.3

Categories of Radioactive Packages

Transport Index (TI) Category	Maximum Radiation Level at Any Point on External Surface	
0	<0.005 mSv/h	I-WHITE
>0, but ≤1	>0.005 ≤ 0.5 mSv/h	II-YELLOW
>1, but ≤10	>0.5 but ≤2 mSv/h	III-YELLOW
10	>2, but not more than 10 mSv/h	III-YELLOW[a]

[a] Shall be transported under "exclusive use."

9.5.2   Transport of Brachytherapy Sources within the Hospital

As a general rule all transport distances should be kept as short as possible. Public lifts should be avoided. Transport containers should be designed to have a surface dose rate less than 100 mSv/h and must be accompanied at all times by appropriately qualified personnel. It is advisable to design transport containers with large wheels, which make it easier to negotiate rough surfaces. Also, when transported within the hospital all sources and containers must be labeled.

Finally, consideration must be given to transport of patients with radioactive sources in situ. This may be the case if transfer from theatre to a CT scanner is required. Again, the shortest possible routes should be taken and public areas avoided. Like in all situations when handling radioactive material, good planning and documentation are essential.

References

1. Cleaves, M. A., Radium: With a preliminary note on radium rays in the treatment of cancer, *Med. Rec.*, 64, 601–6, 1903.
2. Cross, F. H., Miller, H., and Mussel, L. E., Unusual radium accident, *Brit. J. Radiol.*, 24, 122, 1951.
3. Vandivert, V. V., and Holden, F. R., Hazard evaluation and control after spill of 40 mg of radium, *Nucleonics*, 11, 45–47, 1953.
4. Marshall, L. D., Norris, W. P., Gustafson, M., and Speckman, T. W., Transport of radium needles from hospital and its elimination from the human body following dental exposures, *Radiology*, 61, 903–15, 1953.
5. Looney, W. B., and Archer, V. E., Radium inhalation accident: Radium excretion study, *Am. J. Roentgenol. Radium Therapy Nuclear Med.*, 75, 548–55, 1956.
6. Gallaghar, R. G., and Saenger, E. L., Radium contaminations and their associated hazards, *Am. J. Roentgenol. Radium Therapy Nuclear Med.*, 77, 511–23, 1957.
7. Krabbenhoft, K. L., Radium accidents, *Am. J. Roentgenol. Radium Therapy Nuclear Med.*, 83, 584–85, 1965.
8. International Commission on Radiation Units and Measurements, *Dose and Volume Specification for Reporting Intracavitary Therapy in Gynecology*, Report 38, Bethesda, MD: ICRU, 1985.
9. Henschke, U. K., "Afterloading" applicators for radiation therapy of carcinoma of uterus, *Radiology*, 74, 834, 1960.
10. Henschke, U. K., Hilaris, B. S., and Mahan, G. D., Afterloading interstitial and intracavitary radiation therapy, *Am. J. Roentgenol. Radium Therapy Nuclear Med.*, 90, 386–95, 1963.
11. Haie-Meder, C., Potter, R., van Limbergen, E., et al., Recommendations from Gynaecological (GYN) GEC-ESTRO Working Group (I): concepts and terms on 3D image based 3D treatment planning in cervix cancer brachytherapy with emphasis on MRI assessment of GTV and CTV. *Radiotherapy Oncology*, 74,(3), 235–44, 2005.

12. Tod, M. and Meredith, W. J., Treatment of cancer of cervix uteri: A revised 'Manchester method,' *Brit. J. Radiol.*, 26, 252, 1953.

13. Gerbaulet, A., Poter, R., Mazeron, J. J., et al., Eds., *The GEC ESTRO Handbook of Brachytherapy*, ESTRO, Belgium, 2002.

14. Waksman, R., Ed., *Vascular Brachytherapy*, 2nd ed., Futura Publishing Co., New York, 1999.

15. U.S. Nuclear Regulatory Council, Events reported to NRC involving HDR units 1/1990–3/2002, 2002.

16. International Commission on Radiation Protection, Prevention of high-dose-rate brachytherapy accidents, *Annals of ICRP*, 97, 2005.

17. Nath, R., Anderson, L. L., Meli, J. A., et al., Code of practice for brachytherapy physics: Report of the AAPM Radiation Therapy Committee Task Group No. 56, *Med. Phys.*, 24, 1557–98, 1997.

18. U.S. Nuclear Regulatory Council, *Loss of an Iridium-192 Source and Therapy Misadministration at Indiana Regional Cancer Center, Indiana, Pennsylvania, on November 16, 1992*, NUREG-1480, Rockville, MD: U.S. Nuclear Regulatory Council, 1993.

19. International Commission on Radiation Protection, Radiation safety aspects of brachytherapy for prostate cancer using permanently implanted sources, *Annals of the ICRP*, 98, 2005.

20. National Council on Radiation Protection and Measurements, *Structural Design and Evaluation for Medical Use of X-rays and Gamma Rays of Energies up to 10 MeV*, NCRP Report 49, NCRP, 1976.

21. Parry, J. M., Kehoe, T., and Sutton, D. G., Radiotherapy: Brachytherapy, in *Practical Radiation Protection in Health Care*, eds. C. J. Martin and D. G. Sutton, New York: Oxford University Press, 2002, 307–26.

22. Nath, R., Anderson, L. L., Luxton, G., et al., Dosimetry of interstitial brachytherapy sources: Recommendations of the AAPM Radiation Therapy Committee Task Group No. 43, *Med. Phys.*, 22, 209–34, 1995.

23. Rivard, M. J., Coursey, B. M., DeWerd, L. A., et al., Update of AAPM Task Group No. 43 Report: A revised AAPM protocol for brachytherapy dose calculations, *Med. Phys.*, 31, 633–74, 2004.

24. Thomadsen, B. R., *Achieving Quality in Brachytherapy*, Institute of Physics Publishing, Bristol, 2000.

25. IAEA, *Regulations for the Safe Transport of Radioactive Material*, Safety Standards Series TS-R-1, Vienna: IAEA, 2000.

Appendix A: LDR Brachytherapy Planning

Date_____

Patient's name_____ UR #_____

Treatment planning: Fraction #_____

Patient name and UR correct_____

Film details: FID, FFD, view, MF recorded on film_____

Treatment unit correct_____

Source #, calibration date, source strength_____

Date and time of implant/treatment correct_____

Markers, catheters identified correctly_____

Anatomical and patient points identified and marked_____

Indexer lengths verified_____

Catheter #	Length (mm)	IL	Catheter #	Length (mm)	IL
1			2		
3			4		
5			6		
7			8		
9			10		
11			12		
13			14		
15			16		
17			18		

Film orientation_____

Image setup entries correct_____

Reconstruction: Correct reconstruction method_____

No. of catheters verified_____

Catheter lengths acceptable_____

Plane and 3D views OK_____

Active source positions OK_____

Indexer lengths OK_____

Dose points OK_____

Dose distribution_____

Normalization_____

Dose prescription correctly entered_____

Dose to dose points OK_____

Dose to other organs acceptable_____

Dose volume OK_____

Plot magnification OK_____

Isodose distribution planes OK_____

Treatment plan data sent to TCS_____

Treatment plan printout_____

Appendix B: Ir-192 HDR Treatment

Patient's name_____   UR #_____

Dose per fraction_____Gy____No. of fractions_____

	1	2	3	4	5
Date	_/_/	_/_/	_/_/	_/_/	_/_/
Patient's name and UR # correct	_/	_/	_/	_/	_/
Treatment unit and activity correct	_/	_/	_/	_/	_/

Cath. #	Ch. #	IL		Cath. #	Ch. #	IL
1				2		
3				4		
5				6		
7				8		
9				10		
11				12		
13				14		
15				16		
17				18		

Diagram with Catheter Locations and Numbers

	1	2	3	4	5
Source locations and dwell times: Verified	_/	_/	_/	_/	_/
Catheter connections and indexer ring: OK	_/	_/	_/	/	_/
Mobile shields (if any) placed strategically	_/	_/	_/	_/	_/
Interlock door-chains	_/	_/	_/	_/	_/
TV cameras focused on patient/monitors	_/	_/	_/	_/	_/
Emergency container in Tx room	_/	_/	_/	_/	_/

Personnel radiation monitors other than TLD	_/_	_/_	_/_	_/_	_/_
Radiation oncologist and anesthetist available	_/_	_/_	_/_	_/_	_/_
No person other than patient in HDR room	_/_	_/_	_/_	_/_	_/_
Treatment initiated					

Posttreatment

Monitored the treatment room					
Catheter/channels OK					
Treatment summary printed out					

Remarks:

_____

_____

_____

_____

_____

_____

_____

_____

_____

_____

_____

_____

_____

_____

_____

_____

_____

_____

_____

Appendix C: Patient Card

Name of Patient:_____ UR #_____

Date of Birth:_____

Name of Patient: <u>CITIZEN John</u>

Date of Birth: <u>May 6, 1943</u>

Registration #: <u>07/01/0001</u>

Address: <u>1. Nowhere Road</u>
              <u>Eden, State 1000</u>

              <u>UTOPIA</u>

              <u>Tel: (01) 1234 5678</u>

IN CASE OF MEDICAL EMERGENCY CONTACT:

              <u>CRAB CANCER HOSPITAL</u>

              <u>101 ANYWHERE STREET</u>

              <u>HEAVEN, UTOPIA 1001</u>

IN PROSTATE: <u>(01) (1) 1234 5678</u>

THIS PATIENT HAS RADIOACTIVE SEEDS OF IODINE-125 IMPLANTED

No. of Seeds: <u>74</u>

Total Activity: <u>931</u> MBq

Date of implant: <u>May 22, 2007</u>

IGNORE THIS CAUTION AFTER MAY 2009

Appendix D: Emergency Procedures
for the Microselectron HDR Unit

1. The radiation oncologist and the physicist should be wearing the digital radiation monitors.

2. The emergency storage container and long tongs should always be available/handy near the HDR unit in the theatre. Ensure that the wire cutter and other emergency handling equipment are at hand.

3. The radiation survey meter should be switched on before the treatment starts.

4. If during treatment any malfunction is detected depress the INTERRUPT button on the control console and start the stopwatch. If the room monitor indicates radiation level, proceed immediately to step 5, or else enter the room and investigate.

5. *Depress the red EMERGENCY STOP at the control console or at the door.* If the room monitor does not indicate any radiation level, proceed to step 11 or else to step 6.

6. The physicist should enter the room with the radiation survey meter, check for radiation level, and *depress the red emergency switch on the unit panel.* If radiation is not detected proceed to step 11 or else to step 7.

7. Push down on the access panel on top of the unit to access the gold hand-crank. *Turn the gold hand-crank clockwise till it blocks.* If radiation is not detected, proceed to step 11 or else to step 8.

8. Disconnect both ends of the suspected channel transfer tube and monitor. If the source is detected in any transfer tube, move the patient away. Proceed to step 9 or else to step 10.

9. *Place the transfer tube in the emergency storage container. Note the time.* Proceed to step 11.

10. The radiation oncologist should enter the room and *remove the applicators/catheters from the patient and place them in the emergency storage container. Note the time.*

11. Disconnect the patient from all supporting and monitoring systems. The patient can now be moved to the recovery room. Print out the treatment data and note the source recovery time.

Appendix E: Sample Check Sheet: Commissioning/ Annual QA Tests for HDR Treatment Unit

1.   SAFETY CHECKS

1.1   Interlocks. These checks should be performed with the unit set up for test treatment with an appropriate applicator connected.

1.1.1 EMERGENCY STOP. Set about 200 s for treatment time. Initiate the treatment. When the source is in the intended dwell position and the dwell timer has started, press the red EMERGENCY button on the control panel. PASS: The source is immediately retracted. FAILURE: The source fails to retract.

1.1.2 TERMINATE. Reset the unit using the reset key. Initiate the treatment. Once the source has reached the dwell position and the dwell timer has started, press the INTERRUPT button. PASS: The source is retracted. FAILURE: The source fails to retract.

1.1.3 DOOR EMERGENCY SWITCH. Initiate treatment. Once the source has reached the dwell position and the dwell timer has started, press the door emergency switch. PASS: The source retracts. FAILURE: The source fails to retract.

1.1.4 DOOR INTERLOCKS. Initiate treatment. Once the source has reached the dwell position and the dwell timer has started, undo the chain (#1) at the wash basin. PASS: The source is retracted. FAILURE: The source fails to retract. Repeat the procedure for the other two chain barriers also. Initiate the treatment. Once the source has reached the dwell position and the dwell timer has started, push open one of the lead lined doors. PASS: The source is retracted. FAILURE: The source fails to retract. Repeat for the other two door panels also.

1.2   CATHETERS

1.2.1 Catheter connections to TRANSFER TUBES. Connect 18 red transfer tubes to the unit. Initiate treatment. PASS: The check cable fails to run out into any channel. FAILURE: The check cable runs out through any or all channels.

1.2.2 Attach appropriate catheters to the transfer tubes, but do not lock them in place. Initiate treatment. PASS: The check cable/source fails to run out. FAILURE: Check cable/source runs out into any or all catheters. Repeat with other types of transfer tubes.

Remove all catheters. Connect one transfer tube (purple or white). Attach a catheter with an acute loop (radius of curvature less than 10 mm). Initiate treatment after disabling extra check cable run. PASS: After check cable run an error code is displayed and unit refuses to send the source out. FAILURE: After the check cable run, the source is sent out.

1.3   COMMUNICATION EQUIPMENT. Check the functioning of the television monitors and the intercom.

1.4. WARNING LIGHTS (RED). Initiate the treatment. Once the source has reached the dwell position and the dwell timer has started, check whether the red indicator lights above the control panel and the two ceiling lights above the interlocking chains are lit. PASS: All lights are lit. FAILURE: One or more have failed to light. Check the monitor cables and connections and the monitor and repeat the procedure.

1.5   BATTERY CHECK for monitors: Check the battery status of the radiation monitors in use and change them if necessary.

2.   DOSIMETRY CHECKS

2.1   SOURCE POSITIONING

2.1.1 SOURCE POSITIONING (with ruler). Attach the red transfer tube to the check ruler and connect to channel 1. Set indexer length as 845 mm. Set the treatment parameters at dwell positions 1, 5, 9, 13, 17, 21, 25, 29, 33, 37, and 41 and set the dwell times to 3–5 s at each position. Initiate check source drive. Note the position reached by the check cable tip. PASS: The reading should be 7 mm more than the set indexer length. FAILURE: Any other value more than ±1 mm.

Initiate treatment. Note the active dwell positions on the ruler as the source steps in. PASS: The positions are within ± 1 mm. FAILURE: Positions outside this tolerance. (Note: The ruler can be used for IL up to 1,000 mm only.)

2.1.2 SOURCE POSITIONING (with catheters). Tape an X-O'mat film (12 × 10 in.) on the table. Place the autoradiograph jig on it with the lead line away from the film. Insert 6 nos. each of SS needles and Proguides and one bronchial catheter in the jig. Place the 2 mm Perspex over the jig. Tape the catheters onto the table as well as with the jig. Insert the Nucletron dummies in each catheter till they reach the tip. Position the C-arm x-ray unit with tube above the table at approximately 75 cm. Expose the films to 70 Kv, 80 mAs. Without disturbing the setup, remove the dummies and connect appropriate transfer cables to the catheters. Set the treatment parameters for dwell positions as in 2.1.1 for 2–3 s (depending upon the source strength) and appropriate ILs. Initiate treatment. After all source positions are irradiated, process the film and check the source positions, as seen by the autoradiograph, with the dummy positions as seen by the radiograph. PASS: The positions within ±1 mm. FAILURE: Outside this range.

2.1.3 SOURCE POSITIONING (with gynecological applicators). Tape an X-O'mat film on the table. Place the autoradiograph jig on it. Place the gynecological catheters with the tip abutting the lead strip. Tape the applicators. Insert appropriate Nucletron dummy catheters in the catheters. Take a radiograph. Set IL 995. Without disturbing the setup, connect the appropriate transfer cables and initiate a treatment with 2–3 s. Process the film and check source position 1 with the dummy position. PASS: Positions within ±1 mm. FAILURE: Outside the range.

2.2. TIMER

2.2.1 Attach a Proguide to the white transfer tube and connect to the unit. Position the Proguide in the timer-check equipment. Set the treatment time for 0.1 s. Initiate treatment and record the pulse time on the display. Repeat the measurement for 0.2, 0.5, 1.0, 2.0, 5.0, 10.0, 30.0, and 60.0 s. PASS: The timer error should be less than 0.02 s for times above 1.0 s; else FAILURE. Set the same time and repeat at least 5 times. Consistency of timer PASS: if the results agree within 0.5%; FAILURE: if the results are outside 0.5%.

2.2.2 Set any convenient dwell time for dwell position 1. Initiate treatment. The source should retract when dwell time is counted down to zero. PASS: If it does. FAILURE: If it does not.

2.3 DECAY FACTOR

2.3.1 Date and time: Check the current date and time.

Check the current AKR with the table of decay chart. PASS: Agree within second decimal. FAILURE: Not agreeing within second decimal.

NOTE: The battery for the emergency motor is checked at every source change. Other electrical parameters are also checked at every source change. The check cable is changed after 5000 runs.

HDR Treatment Control Unit QA   Date_____

1.   Safety Interlocks

	Passed	Failed	Sign.	Comments
1.1.1 Emergency stop				
1.1.2 Terminate switch				
1.1.3 Emergency door				
1.1.4 Chain 1				
1.1.4 Chain 2				

1.1.4 Chain 3				
1.1.4 Door panel 1				
1.1.4 Door panel 2				
1.1.4 Door panel 3				
1.2.1 Cath. connections				
1.2.2 Cath. locks				
1.2.3 Cath. block				
1.3 Commun. equipment				
1.4 Warning lights				
1.5 Battery check				

2.   Dosimetry Checks

2.1.1   Source Positioning: Ruler Check

	Check Cable Tip	#1	#5	#9	#13	#17	#21	#25	#29	#33	#37	#41
Reading												
Variation												

Test: Pass/Fail   Initials_____

2.1.2   Source Positioning: Radiation Check

	Proguide Catheters						Steel Needles						Bronc.
	1	2	3	4	5	6	1	2	3	4	5	6	
Reading													
Variation													

Test: Pass/Fail   Initials_____

	Rotterdam Uterine	Rotterdam Rt. Ovoid	Rotterdam Lt. Ovoid	MR Uterine	MR Rt. Ovoid	MR Lt. Ovoid
Reading						
Variation						

Test: Pass/Fail   Initials_____

2.2.1   Timer Check

Set Time (s)	0.1	0.2	0.5	1.0	2.0	5.0	10.0	30.0	60.0
Pulse time									
Variation									

Test: Pass/Fail   Initials_____

2.2.1 cont'd. Set time for 2 s.

Pulse time (s)				

Consistency test: Pass/fail

2.2.2 Source Retraction at Dwell Time Countdown to Zero

Pass/Fail   Initials_____

2.3   Clock and Decay

Clock_____   TCS_____
                          PLATO_____

AGREE/DISAGREE.   Initial_____

Current source strength:

                          Decay table _____cGy·m^2·h^{-1}
                          As in TCS _____cGy·m^2·h^{-1}

AGREE/DISAGREE.   Initial_____

2.4   Radiological Survey

Source strength:_____cGy·m²·hr⁻¹

Workload: h/year at 2/3 max. source strength. Use factor ... 1.

Location_____Dose Rate (mSv·h⁻¹)_____Estimated Annual Dose (mSv)_____

1

2

3

4

5

6

7

8

9

10

11

12

10

Outlook and Conclusion

Tomas Kron

CONTENTS

10.1 Will There Be Need for Radiation in the Future?

Better outcomes in health care rely on better diagnosis. Both radiology and nuclear medicine at present play an integral role in standard diagnostic procedures. There have been successes in the development of imaging modalities that do not use ionizing radiation: ultrasound and magnetic resonance imaging (MRI) are excellent examples. However, these imaging modalities typically complement those based on radiation: x-ray contrast is based on electron density and atomic number, while MRI shows proton density and the physical and chemical environment of these protons. Having more than one imaging modality available allows for differential diagnoses. It is therefore not surprising that the number of computed tomography (CT) scans per head of population has increased dramatically over recent years.[1,2] Other aspects that contribute to the increase in medical exposure from diagnostic use of x-rays are developments such as four-dimensional (4D) CT scanning[3] and increasingly the use of interventional radiography.

Nuclear medicine has also seen strong growth over the last years. In particular the use of positron emission tomography (PET) has surged and made a huge impact in oncology.[4,5] In a clinical situation where neoplastic growth arises from normal tissue, it is often difficult to distinguish the tumor from the surrounding normal tissue. The ability to show metabolic activity in addition to anatomy is very helpful in guiding the diagnosis in this case. However, PET does not stop here, and the impact of new tracers such as deoxy-fluorothymidine (FLT; for detection of proliferation) or fluoromisonidazole (FMISO; for detection of hypoxia) will further broaden the spectrum of applications for nuclear medicine.[6]

It is also interesting to note that the use of two imaging modalities in combination as in PET-CT[7,8] or more recently single photon emission computed tomography (SPECT)-CT[9] has had a dramatic impact on patient management. There is no doubt that the ability of ionizing radiation to generate information from inside patients noninvasively will ensure that x-rays and gamma rays will be in use for years to come.

For cancer treatment it is unlikely that radiation therapy will be redundant in the next 20 years. There are many aspects of radiotherapy, which make it a unique contribution to cancer management:

- External-beam radiotherapy is noninvasive.
- The combination of imaging tools and radiotherapy delivery (often referred to as image-guided radiation therapy [IGRT]) has made radiotherapy much more accurate.
- Improvements in delivery technology such as intensity-modulated radiation therapy (IMRT) reduce the toxicity.
- Radiotherapy achieves effective cell kill of many orders of magnitude.
- Radiotherapy is comparatively cheap.
- Radiotherapy delivery is fast and can be uncomplicated and very effective for palliation.

At present about 40% of cancer patients will have radiotherapy at some stage in their management. It is estimated that this number should grow to above 50% to ensure that everyone who could benefit from radiotherapy will actually have it.[10]

Even brachytherapy has been growing in recent years after some years of stagnation. In many hospitals brachytherapy had been phased out because it requires access to an operating theatre, is time consuming for the medical practitioner, and requires extensive training and skills. However, the introduction of new brachytherapy techniques for common cancers such as breast and prostate has recently reversed this trend. In particular, brachytherapy of prostate cancer with radioactive 125-iodine seeds has been a significant success. Patients frequently request brachytherapy because it is perceived to be generally a good sparing method for healthy organs.[11]

Given all the discussion above, it appears certain that the use of radiation for diagnostic and therapeutic purposes is going to stay with us for the foreseeable future—and radiation protection with it.

10.2 New Challenges in Radiation Protection

However, there are new challenges (and opportunities) in radiation protection. These can be roughly classified into three areas:

* Physical and technological
* Radiobiological
* Societal and ethical

10.2.1 Physical and Technological Challenges for Radiation Protection

Medical technology is developing fast—this also applies to both imaging and radiation therapy. The introduction of new, more sensitive image acquisition systems such as computed and digital radiography may in principle reduce the dose to the patient. However, this outcome is not always achieved as radiologists often need to be encouraged to trade off some image quality to achieve the lower radiation dose, even when the quality of the diagnosis is not affected. Faster image acquisition systems also extend the utility of equipment—CT fluoroscopy[12] and 4D CT scanning[13] allow for acquisition of time-resolved three-dimensional image sets. The diagnostic advantage in biopsy taking and assessment of cardiac problems and lung cancers is obvious. However, these procedures typically deliver a higher radiation dose to the patient and risk-benefit analysis must be considered.

Improved nuclear technology and better radiochemistry have allowed the production of radioisotopes with new capabilities in nuclear medicine and brachytherapy. For the former, a larger variety of tracers promises imaging of more physiological processes. In brachytherapy, a higher specific activity allows for miniaturization of radiation sources, resulting in faster and more accurate treatment.

Overall, radiotherapy has changed over the last 10 years. The introduction of intensity-modulated radiation therapy allows more precise delivery of radiation. This is achieved by subdividing large radiation fields into many small segments, with more flexibility in optimizing the dose distribution. This increase in flexibility is "bought" by increasing the overall beam-on time, resulting in more leakage radiation. While no clinical consequences have been observed, there is considerable discussion about the potential of secondary cancers due to adoption of new technology.[14,15] There is no doubt that IMRT has improved radiotherapy delivery, and several clinical trials have demonstrated a reduction in toxicity.[16,17] However, in addition to the knowledge about increased leakage radiation, we often lack the dosimetry tools to assess radiation dose levels at large distances from the primary field, such as the breast dose in a gynecological treatment of the cervix.

Other advances in radiotherapy delivery are related to image guidance. Image-guided radiation therapy uses daily imaging to localize the tumor within the patient. If this is done using CT technology, there may be a significant additional dose to the patient over a course of thirty to forty fractions of radiotherapy. Only time will tell if for some patients the increased risks due to radiation will be larger than the benefits of increased daily accuracy.

The example of IGRT illustrates also another new challenge in radiation medicine: the combination of imaging and treatment modalities. PET-CT is another example where the overall dose to the patient is difficult to assess—it has different sources, different time-dose profiles, and different dose distributions.

10.2.2 Radiobiological Challenges for Radiation Protection

Radiobiological challenges for radiation protection come from at least two different angles: the dramatic developments in modern biology and the need to develop models that can be used for practical radiation protection.

Epidemiological evidence of radiation risks has served us well over the last century. However, there are significant limitations for a system of radiation protection that is based on epidemiological evidence only. These have been discussed in Chapter 3 and include in particular a lack of ability to customize protection to the needs of individuals. The recent advances in molecular biology and genetics provide a unique opportunity to improve the scientific basis for radiation protection. They allow for better understanding of the underlying mechanisms as well as for identification of biological risk factors. It will be a significant challenge for radiation protection professionals to update their knowledge and skills to be able to tap into this resource.

There is an opportunity to learn more about biological effects of radiation in the context of radiotherapy clinical trials, including a translational component. Many radiotherapy trials include the collection of tissue samples with assessment for a variety of genetic factors (compare, e.g., http://www.rtog. org/ or http://www.eortc.be/). Many of them may be linked to radiation sensitivity, and it will be important to use these data to enhance our understanding of radiobiology.

On the other hand, there will always be a need for robust models of radiation effects that can be used in political decision making and day-to-day radiation protection practice. It is important to realize that these models must include simplifications. The linear no-threshold model discussed in Chapter 3 is a good example of this: it does not allow for different tumor types or radiation sensitivity in individuals. However, it appears at present to be a suitable model for the purpose of developing general radiation protection principles. It will remain a challenge to refine this and other models and extend them to scenarios where more specific risk assessments need to be made.

10.2.3 Societal and Ethical Challenges for Radiation Protection

Ethical issues have always featured in the discussion of radiation safety, as demonstrated nicely in two papers in *Health Physics* at the end of the last century.[18,19] There is no doubt that awareness of risks (and the impossibility of avoiding them) is increasing within modern societies. The discussion on screening procedures such as mammography[20] and whole body CT[21] is a good example within medicine, while climate change and nuclear power generation serve to illustrate the political dimension of this debate.

On the other hand, many societies have developed a culture of legal awareness. This has several aspects:

- There is a trend toward ensuring all documentation is legally "tight," including disclaimers and waivers of responsibility.

- The risk of litigation may lead persons to avoid practices that may carry a risk. As this risk and the possible damages awarded to injured persons are reflected in insurance premiums, some practices may become too expensive.

- The risk of legal implications may hinder the reporting of incidence. This reporting and dissemination of information is essential to ensure learning from accidents.[22]

Another important change in radiation protection is the inclusion of the environment and nonhuman species in radiation safety consideration.[23,24] Previously, it had been implicitly assumed that adequate protection of humans would ensure that everything else is adequately protected. It is now recognized that this may not be the case.

In the context of medicine the most profound change has probably been the increased awareness of patients. Many patients are better informed about their illnesses and their rights, and the Internet has provided them with a tremendous amount of readily available information. The need to provide relevant information to patients has also been recognized by the International Commission on Radiological Protection (ICRP). In their document *Radiation and Your Patient: A Guide for Medical Practitioners*, the ICRP provides persons in the medical environment with concise, useful, and up-to-date information on radiation protection.[25] Publications such as this will be very helpful in ensuring patients and their carers are well informed on radiation issues and will enable them to provide the informed consent necessary before many medical procedures may be undertaken.

10.3 Future Trends?

It is beyond the scope of the present book to predict the future. However, two interesting publications may serve as an indicator of how radiation protection may develop over the next years.

10.3.1 Research Needs for Radiation Protection

The National Council on Radiation Protection and Measurements in the United States (NCRP) considered these issues in Report 117, published in 1993.[26] While already more than 10 years old, it still reflects important current trends. On the top of the list is cellular and molecular biology, followed by dose determination and risk assessment. There is also an interesting section on prevention, intervention, and perception—the latter is intrinsically linked to public policy and the need for societies to assess their values in respect to risks, be it from radiation or other sources.

10.3.2 Fundamental Safety Principles

It is apt to conclude this book with a comment on a recent publication[27] on fundamental safety principles that is jointly sponsored by nine international organizations, including the International Atomic Energy Agency (IAEA), the Food and Agricultural Organization of the UN (FAO), the International Labor Organization (ILO), and the World Health Organization (WHO). This document accepts the no-threshold model of radiation risk discussed in Chapter 3 and aims to establish fundamental principles that can guide safety programs. Ten fundamental safety principles are identified. They are an elegant summary of elements that are contained in other documents by the IAEA and the ICRP. They also reflect back on the contents of Chapters 1 and 5 of the present book and as such provide a useful checklist for persons aiming to set up a radiation safety program in medicine:

1. Responsibility for safety
2. Role of government
3. Leadership and management for safety
4. Justification of facilities and activities
5. Optimization of protection
6. Limitation of risks to individuals
7. Protection of present and future generations
8. Prevention of accidents
9. Emergency preparedness and response
10. Protective actions to reduce existing or unregulated radiation risks

The present book aims to assist in this process.

References

1. Mettler, F., Wiest, P., Locken, J., and Kelsey, C., CT scanning: patterns of use and dose, *J. Radiol. Prot.*, 20, 353–59, 2000.
2. Wise, K., and Thomson, J., Changes in CT radiation doses in Australia from 1994 to 2002, *Radiographer*, 51, 81–85, 2004.
3. Vedam, S. S., Keall, P. J., Kini, V. R., Mostafavi, H., Shukla, H. P., and Mohan, R., Acquiring a four-dimensional computed tomography dataset using an external respiratory signal, *Phys. Med. Biol.*, 48, 45–62, 2003.
4. Bomanji, J. B., Costa, D. C., and Ell, P. J., Clinical role of positron emission tomography in oncology, *Lancet Oncol.*, 2, 157–64, 2001.
5. Belhocine, T., Spaepen, K., Dusart, M., Castaigne, C., Muylle, K., Bourgeois, P., Bourgeois, D., Dierickx, L., and Flamen, P., 18FDG PET in oncology: The best and the worst [Review], *Int. J. Oncol.*, 28, 1249–61, 2006.
6. Saleem, A., Potential of PET in oncology and radiotherapy, *Br. J. Radiol.*, Suppl 28, 6–16, 2005.
7. Ell, P. J., The contribution of PET/CT to improved patient management, *Br. J. Radiol.*, 79, 32–36, 2006.
8. von Schulthess, G. K., Steinert, H. C., and Hany, T. F., Integrated PET/CT: Current applications and future directions, *Radiology*, 238, 405–22, 2006.
9. Madsen, M. T., Recent advances in SPECT imaging, *J. Nucl. Med.*, 48, 661–73, 2007.
10. Delaney, G., Jacob, S., Featherstone, C., and Barton, M., The role of radiotherapy in cancer treatment: Estimating optimal utilization from a review of evidence-based clinical guidelines, *Cancer*, 104, pp. 1129–1137, 2005.
11. Cooperberg, M. R., Lubeck, D. P., Meng, M. V., Mehta, S. S., and Carroll, P. R., The changing face of low-risk prostate cancer: Trends in clinical presentation and primary management, *J. Clin. Oncol.*, 22, pp. 2141–2149, 2004.
12. Lucey, B. C., Varghese, J. C., Hochberg, A., Blake, M. A., and Soto, J. A., CT-guided intervention with low radiation dose: Feasibility and experience, *AJR Am. J. Roentgenol.*, 188, 1187–94, 2007.
13. Li, T., Schreibmann, E., Thorndyke, B., Tillman, G., Boyer, A., Koong, A., Goodman, K., and Xing, L., Radiation dose reduction in four-dimensional computed tomography, *Med. Phys.*, 32, 3650–60, 2005.
14. Hall, E. J., Intensity-modulated radiation therapy, protons, and the risk of second cancers, *Int. J. Radiat. Oncol. Biol. Phys.*, 65, 1–7, 2006.
15. Moiseenko, V., Duzenli, C., and Durand, R. E., In vitro study of cell survival following dynamic MLC intensity-modulated radiation therapy dose delivery, *Med. Phys.*, 34, 1514–20, 2007.
16. Kam, M. K., Teo, P. M., Chau, R. M., Cheung, K. Y., Choi, P. H., Kwan, W. H., Leung, S. F., Zee, B., and Chan, A. T., Treatment of nasopharyngeal carcinoma with intensity-modulated radiotherapy: The Hong Kong experience, *Int. J. Radiat. Oncol. Biol. Phys.*, 60, 1440–50, 2004.
17. Zelefsky, M. J., Fuks, Z., Happersett, L., Lee, H. J., Ling, C. C., Burman, C. M., Hunt, M., Wolfe, T., Venkatraman, E. S., Jackson, A., Skwarchuk, M., and Leibel, S. A., Clinical experience with intensity modulated radiation therapy (IMRT) in prostate cancer, *Radiother. Oncol.*, 55, 241–49, 2000.
18. Shrader-Frechette, K., and Persson, L., Ethical issues in radiation protection, *Health Phys.*, 73, 378–82, 1997.
19. Schwarz, D., Ethical issues in radiation protection, continued, *Health Phys.*, 75, 183–86, 1998.

20. Ng, K. H., Jamal, N., and DeWerd, L., Global quality control perspective for the physical and technical aspects of screen-film mammography: Image quality and radiation dose, *Radiat. Prot. Dosimetry*, 121, 445–51, 2006.
21. Prokop, M., Cancer screening with CT: Dose controversy, *Eur. Radiol.*, 15 (Suppl. 4), D55–61, 2005.
22. International Atomic Energy Agency, *Lessons Learned from Accidental Exposures in Radiotherapy*, IAEA Safety Reports Series 17. Vienna: IAEA, 2000.
23. Clarke, R. H., and Holm, L. E., The commission's policy on the environment, *Ann. ICRP*, 33, 201–3, 2003.
24. Clarke, R., 21st century challenges in radiation protection and shielding: Draft 2005 recommendations of ICRP, *Radiat. Prot. Dosimetry*, 115, 10–15, 2005.
25. International Commission on Radiological Protection, *1990 Recommendations of the International Commission on Radiological Protection*, ICRP Report 60. Oxford: Pergamon Press, 1991.
26. National Council on Radiation Protection and Measurements, N. R. 1., *Research Needs for Radiation Protection. Report 117*, Bethesda, MD: NCRP, 1993.
27. International Atomic Energy Agency, *IAEA Safety Standards: Fundamental Safety Principles*. Vienna, Austria: IAEA, 2006.

Glossary

absorbed radiation dose, or absorbed dose: The amount of energy deposited per unit mass. The unit is the gray. See Chapter 2 for more detail.

acceptance testing: Relates to a series of tests carried out after new equipment has been installed or major modifications have been made to existing equipment in order to verify compliance with contractual and manufacturer's specifications.

afterloading applicators (in brachytherapy): Hollow applicators, catheters, or needles can be placed either in or on the tumor volume. At a later time the active sources can be remotely loaded into the applicators as needed, without any radiation hazard to the operating oncologist or associated staff. The loading of the sources can be performed manually or by a remote automatic system.

air kerma strength or air kerma rate (in brachytherapy): Radiation output from any source expressed in $\mu Gy \cdot h^{-1}$ or $cGy \cdot h^{-1}$ at a specified distance (usually 100 cm for HDR brachytherapy sources) along the equatorial axis.

apoptosis: Programmed cell death; a mode of cell disintegration resulting in cell death. It is characteristic for cells that sense they are damaged (e.g., by radiation) and are unable to repair. These cells may undergo apoptosis, a type of cell suicide. The gene p53 is involved in regulating apoptosis.

automatic brightness control (ABC): A technology whereby the image on a video monitor, produced from an image intensifier or flat-panel detector, is maintained at uniform brightness regardless of the anatomy viewed.

automatic exposure control (AEC): A technology whereby an exposure is terminated automatically by the x-ray generator, when the dose to the image receptor reaches a predetermined level.

becquerel (Bq): The SI unit for the rate of radioactive decay (activity) of a radioactive source. One becquerel is defined as one decay per second. See also **curie (Ci)**.

brachytherapy: The mode of radiation therapy where small sealed radiation sources are placed in the tumor tissue, or over it at distances of the order of a centimeter. The sources may be placed temporarily or implanted permanently.

cell cycle: Describes the growth and division of cells. It consists of several phases: G (gap or growths), S (DNA synthesis phase), and M (mitosis or cell division phase).

collective equivalent dose and collective effective dose: Relate to exposures of groups of people. The purpose of these quantities is to predict the total consequences of an exposure of a population, either from a single event or from a long-term situation in an environment. The ICRP states that collective dose is obtained as the sum of all individual doses over a specified time period from a source.

collimator: A fixed or adjustable device to limit the useful x-ray beam to specific dimensions.

committed dose, committed equivalent dose, and committed effective dose: Arise from the intake of radioactive material into the body. In these cases there will be a time period during which the material will be undergoing radioactive decay within the subject committing them to a radiation dose until the material either decays, is excreted from the body, or the subject dies. It is then said that the subject is committed to receive a radiation dose for the coming years, hence the term committed dose.

computed tomography dose index (CTDI): Applied in the context of CT scanning; the integral along a line parallel to the axis of rotation (z) of the absorbed dose profile for a single rotation and a fixed table position divided by the nominal thickness of the x-ray beam. If the integration is restricted to a distance of 100 mm centered on the center of the dose profile it is denoted as $CTDI_{100}$.

controlled area: An area to which access is subject to control and in which employees are required to follow specific procedures aimed at controlling exposure to radiation.

curie (Ci): The older unit of radioactive decay rate (activity) before being replaced by the becquerel (see also **becquerel**). For a radioactive source 1 Ci represents 3.7×10^{10} decays per second (or 37 GBq).

diagnostic reference level (DRL) for medical exposure: A reference level of dose likely to be appropriate for average-sized patients undergoing medical diagnosis and treatment. If a survey of doses indicates substantial departures from DRLs, the cause should be investigated.

DNA (deoxyribonucleic acid): A molecule that contains the genetic information for all living cells. It is essential for functioning, development, and reproduction of cells. In radiation protection DNA is considered the most important target for radiation damage.

dose area product (DAP): The product of the absorbed dose in air and the area of the x-ray beam at a point in a plane perpendicular to the central axis of the x-ray beam. It does not include contributions from backscatter but is a useful dosimetric quantity for fluoroscopic and other complicated radiological examinations. The DAP may be expressed in units of $Gy \cdot cm^2$.

dose constraint: The dose received by an individual from a single source. This is usually associated with controllable situations or emergencies.

dose/dose rate effectiveness (DDREF) factor: Used to describe the difference between radiation effects observed at high radiation doses and dose rates and effects expected at low doses and low dose rates. DDREF is typically chosen to be 2 for radiation protection purposes. It is part of the LNT model.

dose length product (DLP): A dosimetric quantity applicable to a complete CT examination; is defined for sequential scanning in terms of the normalized weighted CTDI through the equation

$$\text{DLP} = \sum_i {}_n\text{CTDI}_w \, T \, N \, Q$$

where N is the number of slices in a given sequence, each of thickness T acquired with a current time product Q, and the summation is over all scan sequences forming part of the examination. For helical scanning the equation may be rewritten as

$$\text{DLP} = \sum_i {}_n\text{CTDI}_w \, T \, A \, t$$

where T is the nominal irradiated beam width, A is the x-ray tube current, and t is the total acquisition time for each sequence. ${}_n\text{CTDI}_w$ is determined for a single slice as in sequential scanning. DLP is usually expressed in units of mGy·cm.

dose limit: The dose received by an individual from all regulated sources under normal circumstances.

effective dose: Used to allow for a variation in the sensitivity of tissue types to radiation. For example, gonads are much more sensitive to radiation than skin. The result of this is the addition of a tissue weighting factor w_T, which operates as a multiplier of the equivalent dose to obtain the effective dose, E. For a more detailed discussion see Chapter 2.

electronvolt (eV): A unit of energy and the amount of kinetic energy that an unbound electron will gain when passing through an electric field of 1 V within a vacuum (1.602×10^{-19} J). When discussing the energy of radiation, the electronvolt is used more commonly than joules.

entrance surface dose (ESD): The value of the absorbed dose in air, including backscatter, at the intersection of the central axis of the x-ray beam with the entrance surface of the patient. It is a useful dosimetric quantity for simple radiological examinations and is usually expressed in units of mGy.

entrance surface dose air kerma (ESAK): The KERMA (see **KERMA**) measured free in air at the entrance surface to the patient.

equivalent dose: Term used to describe the differing effects of different types and energies of radiation on the same tissue. For example, most tissue is much more sensitive to radiation in the form of heavy charged particles such as alpha radiation than it is to electromagnetic radiation such as gamma radiation or x-rays. To calculate the equivalent dose from a particular radiation type, a multiplier called a radiation weighting factor (w_R) is introduced to convert absorbed dose to equivalent dose, H_T. See Chapter 2 for more details.

exposure: Calculated in terms of the amount of ionization per mass of air by x-rays or gamma rays up to energies of around 3 MeV. At higher energies the increasing range of secondary electrons make measurement of exposure at a point impractical. See Chapter 2 for more details.

filtration: Modifies the spectral distribution of an x-ray beam as it passes through the filter material, due to the preferential attenuation of particular photon energies in the radiation beam.

added filtration: Quantity indicating the filtration effected by added filters in the useful beam, but excluding inherent filtration.

inherent filtration: The filtration effected by the irremovable materials of the x-ray tube assembly through which the radiation beam passes before emerging from the x-ray tube assembly.

total filtration: The total of inherent filtration and added filtration between the radiation source and the patient or a defined plane.

flat panel detector: A transducer that employs solid-state technology to convert an x-ray image to an electronic image. The image may be viewed using a video chain, as in fluoroscopy, or on a computer monitor when used as an alternative to film-screen technology.

half-value layer (HVL): The thickness of a sheet of given material required to reduce the intensity of a particular radiation beam to half of its original intensity. The HVL is dependent on both the energy of the beam and the absorbing material. See also tenth value layer.

HDR brachytherapy: Brachytherapy treatments where the dose rate to the target tissue is >12 Gy/h. Generally the source strength will be 200–500 GBq. At this dose rate, the radiobiological response is a function of the dose per fraction and not the dose rate, as the treatment time for each fraction will be the order of a few minutes (as with external beam radiotherapy). HDR brachytherapy can be performed only with a remote afterloading unit.

hormesis: Term used for beneficial effects of radiation at low doses.

hypoxia: The state of cells with low oxygen tension. As oxygen is involved in some radiation damage, hypoxia tends to make cells more radioresistant.

ICRP: International Commission on Radiological Protection.

ICRU: International Commission on Radiation Units and Measurements, Inc.

image intensifier: A transducer that employs vacuum tube technology to convert an x-ray image to a light image suitable for viewing by a video chain.

incident (radiation): Any unintended or ill-advised event when using ionizing radiation apparatus, specified types of nonionizing radiation apparatus, or radioactive substances that results in, or has the potential to result in, an exposure to radiation to any person or the environment, outside the range of that normally expected for a particular practice, including events resulting from operator error, equipment failure, or the failure of management systems that warranted investigation (from Australian Radiation Protection and Nuclear Safety, *Authority Draft Code of Practice: Radiation Protection in the Medical Applications of Ionizing Radiation,* 2007).

indexer length (IL) (in brachytherapy): In HDR brachytherapy units, the source can be driven out from a minimum to a maximum as given by the type of unit. However, the maximum drive out in a specific applicator or catheter will depend upon its length, corresponding to the possible first source dwell position in that catheter. This is known as the indexer length for that applicator/catheter.

interstitial implants (in brachytherapy): This is the method in which brachytherapy sources are either implanted into a target tissue and left in situ permanently or placed within an applicator, catheter, or needle for a specific time. The former technique is known as *permanent implants* and the latter as *temporary implants.*

interventional radiology: Procedures are guided therapeutic and diagnostic interventions, by percutaneous or other access, usually performed under local anesthesia or sedation, with fluoroscopic imaging used to localize the lesion/treatment site, monitor the procedure, and control and document the therapy.

intracavitary brachytherapy applications: Tumors in cavities at body sites such as the uterine cervix can be treated by inserting brachytherapy sources in applicators placed inside the cavity.

intraluminal brachytherapy applications: Tumors may grow around the wall of certain organs such as the esophagus, bronchus and its branches, bile duct, and so on. Brachytherapy applicators with sources can be placed in the lumen of these organs to irradiate an annular volume.

intraoperative brachytherapy: After excision of a tumor mass, the remaining tumor tissue or tumor bed can be irradiated intraoperatively or perioperatively. In the intraoperative procedure, a mould is placed over the tissue and irradiated, or alternately, an appropriate applicator is placed in position and the source can be placed into this applicator to give a single irradiation. Mostly this method uses only a single fraction. In the perioperative technique, the applicators or catheters are placed surgically and the incision closed. Later, after due treatment planning, the treatments are effected by multiple fractions using manual or remote afterloading systems. After completion of the treatment the applicators/catheters are removed.

KERMA: Acronym for kinetic energy released in matter. Defined as the amount of energy transferred from photons to kinetic energy of electrons.

LDR brachytherapy: Performed with sources with relatively low strength, of about 1–2 GBq. The dose rate to target tissue will be 0.4–2 Gy/h, and this is known as low-dose-rate (LDR) brachytherapy. In the early days of brachytherapy all treatments were LDR brachytherapy since available radium sources had lower strengths. The radiobiological response of a specific tissue is a function of the dose rate and total treatment time.

lead equivalent: At a specified kVp and x-ray beam quality, the thickness of lead effecting the same attenuation as the material under consideration.

linear energy transfer (LET): The amount of energy deposited along the track of an ionizing particle. LET is typically given in keV/μm.

linear no-threshold (LNT) model: Forms the basics of risk estimation in radiation protection. It assumes that radiation events observed at high radiation doses can be extrapolated to small doses (including a *DDREF factor*).

Manchester Point A: Term used in brachytherapy. In the radium dosage system evolved in Christie Hospital and Radium Institute, Manchester, United Kingdom, an anatomical point was defined in the cervical uteri to prescribe a maximum tolerance dose. The point defined was the point at which, in most anatomy of women, the ureter crossed the uterine artery. In an x-ray image, point A is located 2 cm superior to the external os along the midline of the uterus and 2 cm lateral to it. This point is known as Point A, and in practice this became the point for prescription for cancer of cervical uteri.

medical exposure: The "exposure incurred by patients as part of their own medical or dental diagnosis or treatment; by persons, other than those occupationally exposed, knowingly while voluntarily helping in the support and comfort of patients; and by volunteers in a programme of biomedical research involving their exposure" (IAEA, *Basic Safety Standards*, 1996).

microdosimetry: The discipline studying the distribution of radiation effects on a small scale. As the deposition pattern will affect the radiobiological effectiveness, this is linked to biological outcomes.

mould applications: Many superficial lesions can be treated by placing brachytherapy sources at a desired distance from the skin. The sources are fixed on a material of required thickness, on the side away from the treatment surface. The material is tissue equivalent and pliable so that it can be moulded around the treatment surface. The time for which the patient wears the mould each day depends upon the site of the lesion. These moulds can also be of preloaded or afterloading type.

normalized weighted computed tomography dose index ($_n$CTDI$_w$): The quotient of the weighted CTDI (CTDI$_w$) and the mAs. The $_n$CTDI$_w$ is usually expressed in units of mGy·mAs^{-1}.

occupancy factor (T): The fraction of time that an individual is expected to spend in an irradiated area. It is used for calculation of radiation dose to persons when designing radiation shielding. Guidelines are that areas of full occupancy be assigned T = 1 (occupied offices, controlled areas, laboratories, shops, etc.), partial occupancy be assigned T = 1/4 (corridors, rest rooms, etc.), and occasionally occupied areas be assigned T = 1/16 (car parks, pedestrian areas, toilets, etc). In the absence of site-specific occupancy data, NCRP Report 147 provides suggested values.

occupational exposures: All exposures of workers incurred in the course of their work.

operator: A person who is entitled, in accordance with the responsible person's procedures, to carry out all practical aspects of an exposure.

oxygen enhancement ratio (OER): The ratio between cell kill observed in the presence of oxygen and the cell kill in a hypoxic environment.

pitch: In the context of helical CT scanning, may be defined as the ratio of the patient couch advance per rotation to the total width of the collimated x-ray beam at the patient isocenter.

preloaded applicators: In the early period of brachytherapy, the applicators inserted in a patient had radiation sources already sterilized and loaded. For implants, radioactive needles were directly implanted into the tissue.

public exposure: The "exposure incurred by members of the public from radiation sources, excluding any occupational or medical exposure and the normal local natural background radiation but including exposure from authorized sources and practices and from intervention situations" (IAEA, *Basic Safety Standards*, 1996).

quality assurance: "All those planned and systematic actions necessary to provide confidence that a product or service will satisfy given requirements for quality" (ISO 9000).

reference air kerma rate (RAKR): A quantity used in brachytherapy to describe the radiation output of a specific type of source of a specific radionuclide, expressed in $\mu Gy \cdot h^{-1} \cdot MBq^{-1}$ at 1 cm along the equatorial axis of the source.

referrer: A registered medical practitioner, dentist, or other health professional who is entitled, by the regulatory authority, to refer individuals for medical exposure.

relative biological effectiveness (RBE): The ratio between the dose required for a certain biological effect using reference radiation (usually 60-Co gamma rays) and the dose required for the same effect using the test radiation (e.g., with a different LET).

remote afterloading (in brachytherapy): A method of source loading for both LDR and HDR treatments. The sources in appropriate configuration can be sent into the applicator/catheter through a connecting conduit from a storage device into the applicator/catheter. The device can be programmed for position and dwell time at these locations and withdrawal of the source at the end of the designated time.

tenth-value layer (TVL): The thickness of a sheet of given material required to reduce the intensity of a particular radiation beam to one-tenth of its original intensity. The TVL is dependent on both the energy of the beam and the absorbing material. See also **half-value layer**.

transfer tubes or cables: The conduits through which brachytherapy sources are driven from the treatment unit to the applicator/catheter placed in the patient. The transfer tubes for different types of applicators/catheters are different, and it may not be possible to interchange them.

transport index (TI): For international transport of radioactive material, the package and labeling must conform to the IAEA/IATA regulations, which require the value of TI to be entered on the label. TI is the numerical value of maximum dose rate at the surface of the package in $mSv \cdot h^{-1}$.

TVL: See **tenth-value layer**.

use factor (U): The fraction of total workload in which a radiation source is directed at a particular location. It is mainly used for the purpose of calculating radiation dose when designing a radiation shield. For example, if 75% of radiographs from an x-ray machine are directed vertically downwards and 25% are directed to a chest bucky on a wall, then the use factor for the floor directly under the machine would be 0.75, and the wall behind the chest bucky would be 0.25.

weighted computed tomography dose index ($CTDI_w$): The CTDI obtained by measuring the $CTDI_{100}$ in cylindrical polymethylmethacrylate phantoms and weighting the results according to the following formula:

$$CTDI_w = 1/3 \, CTDI_{100,c} + 2/3 \, CTDI_{100,p}$$

where $CTDI_{100,c}$ refers to the CTDI on the central axis of the phantom and $CTDI_{100,p}$ represents an average of measurements at four different locations 10 mm below the surface around the periphery of the phantom. The $CTDI_w$ is usually expressed in units of mGy.

workload (W): The amount of radiation, of radiation producing equipment, produced in a specified time frame. In diagnostic radiology it refers to the tube current multiplied by beam-on time and is usually expressed in terms of $mA \cdot min$ per week (see Chapter 6). In external beam radiotherapy it generally refers to the total beam-on time per year calculated from the expected total monitor units delivered in a year (see Chapter 8 for a sample calculation).

Index

Bunker, 179
Bystander effect, 48

C

Cancer(s)
  hallmarks of, 52–54
  secondary, 56
Cardiology, 130–132
Cell cycle, 41, 42, 45, 263
Cell kill, models of, 61
Characteristic radiation, 100
Chemical elements, description of, 21
Chernobyl nuclear accident, 55, 56
Classical electron radius, 26
Cobalt-60 units, 174
Collective dose, 8
Collective equivalent dose, 36, 263
Collimator, 264
Committed dose, 35, 264
Compton scattering, 26, 72, 126, 179
Computed radiography (CR), 110
Computed tomography (CT), 4
  automatic exposure control in, 107
  dose index (CTDI), 264
  effective mAs, 115, 116, 118
  installations, use factor for, 135
  overbeaming, 118
  overranging, 117
  pitch, 116, 119
  scanners, multidetector, 116
  simulators, 176
Continuous slowing down
      approximation (CSDA), 28
Controlled area, 88, 134, 264
Corpse, radioactive, disposal of, 163
Coulomb force, 98
  neutrons and, 30
  variation of, 100
CR, *see* Computed radiography
CSDA, *see* Continuous slowing down
      approximation
CT, *see* Computed tomography
CTDI, *see* Computed tomography dose
      index
CTDIw, *see* Weighted computed
      tomography dose index

D

DAP, *see* Dose area product
DDREF factor, *see* Dose/dose rate
      effectiveness factor
Decay
  alpha, 23, 24
  beta, 23, 24
  constant, 22
  rate, 21–22
Deoxyribonucleic acid (DNA), 39
  damage
    fashions of, 39
    repair of, 38
  definition of, 264
  ionizing event in, 61
  schematic diagram of, 40
Detector
  chemical, 79
  film, 78
  gas filled, 72
  neutron, 79
  scintillation, 74, 77
  semiconductor, 77
  thermoluminescent dosimeter, 77
Deterministic effect, 38, 48, 49, 115
Diagnostic and interventional
      radiology, 97–144
  patient protection, 102–128
    computed radiography and digital
        radiography, 110
    computed tomography, 115–119
    diagnostic reference levels,
        108–109
    dose calculations for patients,
        123–128
    fluoroscopy, 110–113
    general and dental radiography,
        110
    high-risk groups, 120–123
    interventional radiology, 113–115
    justification, 102–103
    optimization, 103–107
    quality assurance, 107–108
  protection of occupationally exposed
      individuals and the public,
      128–133
    cardiology and interventional
        radiology, 130–132
    CT fluoroscopy 132–133

For Product Safety Concerns and Information please contact
our EU representative GPSR@taylorandfrancis.com Taylor & Francis
Verlag GmbH, Kaufingerstraße 24, 80331 München, Germany